PALEOECOLOGICAL RESEARCH ON EASTER ISLAND

PALEOECOLOGICAL RESEARCH ON EASTER ISLAND

Insights on Settlement, Climate Changes, Deforestation and Cultural Shifts

VALENTÍ RULL

Institute of Earth Sciences Jaume Almera (ICTJA-CSIC), Barcelona, Spain

ELSEVIER

Elsevier
Radarweg 29, PO Box 211, 1000 AE Amsterdam, Netherlands
The Boulevard, Langford Lane, Kidlington, Oxford OX5 1GB, United Kingdom
50 Hampshire Street, 5th Floor, Cambridge, MA 02139, United States

Notices
Knowledge and best practice in this field are constantly changing. As new research and experience broaden our understanding, changes in research methods, professional practices, or medical treatment may become necessary.

Practitioners and researchers must always rely on their own experience and knowledge in evaluating and using any information, methods, compounds, or experiments described herein. In using such information or methods they should be mindful of their own safety and the safety of others, including parties for whom they have a professional responsibility.

To the fullest extent of the law, neither the Publisher nor the authors, contributors, or editors, assume any liability for any injury and/or damage to persons or property as a matter of products liability, negligence or otherwise, or from any use or operation of any methods, products, instructions, or ideas contained in the material herein.

Library of Congress Cataloging-in-Publication Data
A catalog record for this book is available from the Library of Congress

British Library Cataloguing-in-Publication Data
A catalogue record for this book is available from the British Library

ISBN: 978-0-12-822727-5

For information on all Elsevier publications
visit our website at https://www.elsevier.com/books-and-journals

Publisher: Joe Hayton
Acquisitions Editor: Candice Janco
Editorial Project Manager: Naomi Robertson
Production Project Manager: Sruthi Satheesh
Cover Designer: Christian J. Bilbow

Typeset by SPi Global, India

Working together
to grow libraries in
developing countries

www.elsevier.com • www.bookaid.org

Contents

Acknowledgments   vii
Rapanui glossary   ix
Introduction   xiii

1. The island at present   **1**
  1.1 Geography and geology   1
  1.2 Climate and hydrology   5
  1.3 Soils and land use   11
  1.4 Flora and vegetation   12
  1.5 Archaeological heritage: An outdoor museum   27
  1.6 Conservation: The Rapa Nui National Park   33

2. The prehistory: From human settlement to European contact   **41**
  2.1 From exploration to research   42
  2.2 Prehistoric chronology   44
  2.3 The first settlers   49
  2.4 The ancient Rapanui society   60
  2.5 A narrative of human determinism   87

3. Introduction to Easter Island's paleoecology: Why, where, and how?   **89**
  3.1 Why study paleoecology?   90
  3.2 Initial proposals of prehistoric climate change on Easter Island   95
  3.3 Coring sites   100
  3.4 Cores retrieved and main proxies studied   103
  3.5 Chronology and sedimentary patterns   110
  3.6 Other paleoecological archives   116
  Appendix 3.1   117

4. Paleoecological pioneers: The rising of the ecocidal paradigm   **137**
  4.1 Before paleoecology   138
  4.2 The first systematic pollen analyses   139
  4.3 The first paleoecological synthesis   140
  4.4 The first socioecological synthesis   153
  4.5 Some insights on the pioneer works   158
  Appendix 4.1   163

5. **The transitional phase: Paleoecological impasse**   **169**
    5.1 Rano Kao and the dating problem   169
    5.2 Rano Raraku   172
    5.3 Rano Aroi: More dating problems   178
    5.4 Other studies   178
    5.5 Paleoecological impasse   181

6. **The revival: An opportunity for climate change**   **183**
    6.1 Coring intensification and reanalysis   183
    6.2 Publication resurgence   195

7. **Contributions of paleoecology to Easter Island's prehistory** **217**
    7.1 Discovery and settlement   217
    7.2 Climate changes of the last millennia   222
    7.3 Spatiotemporal deforestation patterns   225
    7.4 Cultural aspects   229

8. **From human determinism to environmental, ecological,
    and social complexity**   **237**
    8.1 The EHLFS approach   237

Epilogue: A plea for true interdisciplinarity   *249*
References   *251*
Species index   *271*
Subject index   *275*

Acknowledgments

The author's research on Easter Island was funded by the Spanish Ministry of Education and Science, through the project GEOBILA (CGL2007-60932/BTE; Principal Investigator: A. Sáez, University of Barcelona, Spain), and was developed in the Laboratory of Paleoecology of the Botanic Institute of Barcelona (2008–2014) and the Institute of Earth Sciences Jaume Almera (2015–2019), of the Higher Spanish Council for Scientific Research (CSIC). The National Forestry Commission (CONAF) of the Chilean Ministry of Agriculture provided the fieldwork permits and logistic assistance when necessary. The Riroroko family, from Hanga Roa, helped with accommodation, transport, storage of coring devices and samples, as well as fieldwork assistance and information about Easter Island's life and traditions.

Discussions with the team members of the abovementioned project are greatly acknowledged, especially Núria Cañellas-Boltà, Santiago Giralt, Olga Margalef, Sergi Pla-Rabes, and Alberto Sáez. Other authors and coauthors of publications on Easter Island's paleoecology with the participation of the author are also acknowledged, including José Álvarez-Gómez, Roberto Bao, Hilary Birks, John Birks, Maarten Blaauw, Raymond Bradley, Teresa Buchaca, Isabel Cacho, William D'Andrea, Adelina Geyer, Armand Hernández, Hans Joosten, Encarni Montoya, Ana Moreno, Leopoldo Pena, Josep Peñuelas, Matthew Prebble, Juan José Pueyo, Jordi Sardans, Irantzu Seco, and Blas Valero-Garcés. However, the views expressed herein are the sole responsibility of the author. The author also acknowledges the comments of Maarten Blaauw, Peter Kershaw, and Andreas Mieth, who reviewed the book proposal for Elsevier. Ann M. Altman, Núria Cañellas-Boltà, Santiago Giralt, Olga Margalef, Shawn McLaughlin, Andreas Mieth, Ana Moreno, and Alberto Sáez are acknowledged for providing images for publication.

This book is dedicated to the memory of the British palynologist John R. Flenley, the pioneer of paleoecological research on Easter Island.

John Roger Flenley (1936–2018)
https://www.nzgs.co.nz/

Rapanui glossary

Brief glossary of Rapanui terms used in this book, following Zizka (1991), McLaughlin (2007), and Edwards and Edwards (2013).

Name	Field	Meaning
Ao	Art	Ceremonial paddle that symbolizes authority or power
Ahu	Architecture	Raised rectangular ceremonial platform
Ariki Mau	Sociopolitics	Paramount chief, from the lineage of Hotu Matu'a within the Miru clan
Ariki paka	Sociopolitics	Noble title for the chiefs of lineages or clans who had kinship ties to the Miru clan
Hare moa	Architecture	Stone house for chicken
Hare paenga	Architecture	Elliptic house resembling an overturned boat reserved for the chiefs
Hau hau	Botany	*Triumfetta semitriloba* (Tiliaceae)
Here hoi	Botany	*Sporobolus africanus* (Poaceae)
Herepo	Botany	*Tetragonia tetragonoides* (Aizoaceae)
Heriki hare	Botany	*Axonopus paschalis* (Poaceae) and *Paspalum forsterianum* (Poaaceae)
Hikukio'e	Botany	Name given to several Cyperaceae species (*Cyperus cyperoides, Cyperus eragrostis, Kyllinga brevifolia, Pycreus polystachyos*)
Hotu Matu'a	Tradition	The legendary founder of the Rapanui culture of Easter Island
Hotu Iti	Sociopolitics	*See Mata Iti*
Ipu kaha	Botany	*Lagenaria siceraria* (Cucurbitaceae)—bottle gourd
Ivi atua	Sociopolitics	Shaman-like priests
Kumara	Botany	*Ipomoea batatas* (Convolvulaceae)—sweet potato
Maika	Botany	*Musa* sp. (Musaceae)—banana
Make Make	Tradition	The Rapanui creator god, forefather, and protector of the Miru clan
Makoi	Botany	*Thespesia populnea* (Malvaceae)
Mahute	Botany	*Broussonetia papyrifera* (Moraceae)—paper mulberry
Mana	Tradition	Spiritual power
Manavai	Architecture	Circular stone shelter for growing plants

(Continued)

Name	Field	Meaning
Manutara	Zoology	*Onychoprion fuscatus* (Charadriiformes: Laridae)—sooty tern
Marikuru	Botany	*Sapindus saponaria* (Sapindaceae)
Mata	Sociopolitics	Clan
Mata'a	Art	Obsidian lithic flakes that have been considered as spear points or tools for domestic labors
Mata Iti	Sociopolitics	Group of clans of lower status, which inhabited the eastern part of the island (also known as Hotu Iti)
Mata Nui	Sociopolitics	Group of clans of higher status, which inhabited the western part of the island (also known as Tuu)
Matato'a	Sociopolitics	Distinguished warrior
Mati	Botany	*Cynodon dactylon* (Poaceae)
Moai	Art	Monolithic stone statues in human form
Moai kavakava	Art	Wooden statuettes representing emaciated men with protruding ribs
Miru	Sociopolitics	The most important Rapanui clan, descendant from Hotu Matu'a and Make Make
Naoho, Ngaoho	Botany	*Caesalpinia major* (Fabaceae)
Ngaatu	Botany	*Scirpus californicus* (Cyperaceae)
Pato	Botany	*Euphorbia serpens* (Euphorbiaceae)
Pipi horeko	Architecture	Stone markers that often indicated places that were prohibited or of restricted access
Poporo	Botany	*Solanum forsteri* (Solanaceae)
Pua	Botany	*Curcuma longa* (Zingiberaceae)
Pua nako nako	Botany	*Lycium carolinianum* (Solanaceae)
Pukao	Art	Hat-like cylindrical headdress for moai made of red scoria
Puringa	Botany	*Verbena litoralis* (Verbenaceae)
Rano	Geography	Volcanic crater with a lake or a swamp inside
Rongorongo	Art	Wood tablets with undeciphered hieroglyphic characters
Rona	Art	Petroglyphs
Taheta	Architecture	Carved or uncarved small stone basins for collecting rainwater
Tangata Manu	Sociopolitics	The Birdman, the representative of Make Make on Earth and maximum authority of Easter Island
Tanoa	Botany	*Ipomoea pes-caprae* (Convolvulaceae)

Name	Field	Meaning
Tapu	Tradition	Sacred, prohibited (enforced by mana)
Taro	Botany	*Colocasia esculenta* (Araceae)
Tavari	Botany	*Polygonum acuminatum* (Polygonaceae)
Te Pito O Te Henua	Tradition	"Navel of the World" or "Land's End," the first name of Easter Island according to the Rapanui tradition
Ti	Botany	*Cordyline fruticosa* (Asparagaceae)
Toa	Botany	*Saccharum officinarum* (Poaceae)
Toki	Art	Stone pick or adze made of basalt used for moai carving
Toromiro	Botany	*Sophora toromiro* (Fabaceae)
Tuava	Botany	*Psidium guajava* (Myrtaceae)
Tuere heu	Botany	*Agrostis avenaceae* (Poaceae)
Tuu	Sociopolitics	See Mata Nui
Umu	Architecture	Earth oven that uses hot stones for cooking food
Yam	Botany	*Dioscorea alata* (Dioscoreaceae)

Introduction

Easter Island (Rapa Nui) has been considered one of the most enigmatic places on Earth for several reasons. First, it is the most remote inhabited island of the planet and is in an intermediate position between South America and the easternmost Polynesian archipelagos (Fig. 1). This has led to wonder about the time of its discovery and settlement, and about the geographical and cultural origin of its first settlers. The present cultural traits of the island and its aboriginal inhabitants (the Rapanui) are clearly of Polynesian origin but hypotheses about possible ancient Amerindian (or Native American) colonization and/or cultural influence have been raised through history (Heyerdahl, 1952; Thorsby, 2016). Regarding its settlement time, a variety of hypotheses exist ranging from about 400 CE (formerly AD) to 1300 CE (Kirch, 2010; Wilmshurst et al., 2011).

Second, the most iconic cultural manifestations of the island are the nearly 1000 megalithic (up to 10 m high and nearly 90 tons of weight) anthropomorphic cult statues known as moai (Fig. 2) erected by the ancient Rapanui culture (Van Tilburg, 1994) (See also the video Rapanui 2008 at https://youtu.be/zTbW4kb-DOU). This ancient society is thought to have flourished well before the European contact (1722), as the Rapanui society that encountered the first Europeans who arrived in the island was considered demographically and technologically unable to develop the moai industry (Flenley and Bahn, 2003; Fischer, 2005). The most traditional enigmas around the moai have been the religious meaning of these statues, the sculpting technology, and the transportation means from their quarry to their final emplacement on special stone altars known as ahu, which are scattered across the whole island (Hunt and Lipo, 2011).

Third, Easter Island has become famous for the assumedly self-provoked cultural collapse of the ancient Rapanui culture that erected the moai, owing to the overexploitation of natural resources, as manifested in total deforestation of the island in a few centuries (Mulloy, 1974; Bahn and Flenley, 1992). This socioecological demise has been considered an ecocide (Diamond, 2005) and has been—and still is—used as a microcosmic model for the whole Earth if current global exploitation practices continue. Other potential deforestation agents would have been human-introduced Polynesian rats, which actively ate fruits of palm trees, thus preventing forest regeneration (Dransfield et al., 1984). Some authors believe that rats were

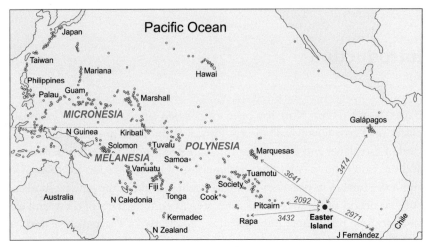

Fig. 1 Sketch map of the Pacific archipelagos. Easter Island is highlighted by a *red* dot and the distances (in km) to the nearest Polynesian and American islands are indicated, according to Flenley and Bahn (2003).

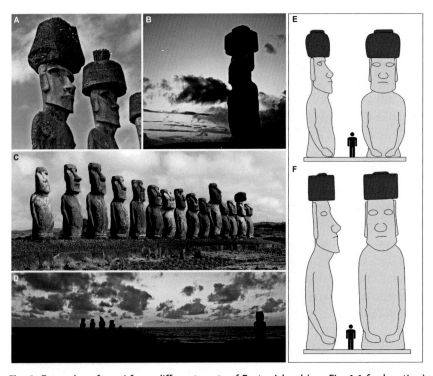

Fig. 2 Examples of moai from different parts of Easter Island (see Fig. 1.1 for location) and size comparison at human scale. Left: (A) Ahu Nau Nau, (B) Ahu Ko Te Riku, (C) Ahu Tongariki, and (D) Ahu Vai Uri and Ahu Tahai. Right: (E) Moai from Ahu Tongariki, (F) Moai from Ahu Te Pito Kura. *(Photos: N. Cañellas (A and C) and V. Rull (B and D). Drawings: (E and F) redrawn and modified from http://www.bradshawfoundation.com/easter/index.php.)*

the main deforestation force and that forest clearing was not the cause of the Rapanui cultural collapse, which did not arrive until after European contact and was actually a genocide (Peiser, 2005; Hunt, 2006). Finally, it has also been discussed whether climate shifts, especially droughts, could have influenced deforestation and/or cultural developments of the ancient Rapanui society (McCall, 1993; Orliac and Orliac, 1998).

For the above reasons, and others, Easter Island has become famous not only in scientific but also in the popular media, which has fostered the appearance of many books, articles, press releases, documentaries, films, and websites with varied tendencies, including magic, obscurantism, science fiction, and similar approaches. Scientific (and pseudoscientific) outcomes are also numerous, especially in the form of books and articles. Therefore, it could be asked whether a new scientific book on Easter Island is really required. The answer is yes, provided it is more than a repetition of what is already published in other books (which is not uncommon) and adds value to the study of Easter Island, in the form of new data, novel interpretations, unconventional perspectives, and/or holistic frameworks accounting for multidisciplinary evidence.

1. Aims and scope

Ethnography, anthropology, and archaeology have dominated Easter Island's research since the early 20th century (Routledge, 1919; Métraux, 1940), which is easy to understand if we consider that most enigmas mentioned above involve human features and activities. Paleoecological research—i.e., the reconstruction of past ecological changes and their natural and anthropogenic drivers—has only acquired relevance on Easter Island during the last 3–4 decades, since the first works of John Flenley and his coworkers (Flenley and King, 1984; Flenley et al., 1991). In spite of their relatively recent inception, these studies have already provided the necessary context for a better understanding of the relationships between environmental/ecological shifts and cultural developments or, in other words, of the complexity of environmental-ecological-cultural systems as functional units. However, this advantage has not been fully exploited. Such a synthetic approach is only possible if any type of determinism, either human or environmental, is circumvented and all the interdisciplinary evidence is considered altogether (Rull, 2018). Due to the lack of truly multidisciplinary initiatives, paleoecological knowledge is still to be fully incorporated into Easter Island's research.

Another reason for this situation may be that paleoecological knowledge on Easter Island is still dispersed across many articles, and an updated and reasoned synthesis is not yet available. Also, a number of paleoecological publications may be too specialized for practitioners of other disciplines and, therefore, difficult to incorporate into their respective working frameworks. Finally, some paleoecological papers have been published in the so-called gray literature or in meeting proceedings and may not be easy to find. This book is a first trial to synthesize and organize all the existing paleoecological knowledge relative to Easter Island in an affordable manner, to facilitate its use in different fields of research. To achieve this, the book provides and analyzes in detail all paleoecological information available to date, organized in a historical manner, to follow its evolution and to highlight the main findings.

2. Book organization

The book is organized into eight chapters. The description of the island (Chapter 1) is succinct as the main aim is not providing a thorough characterization but to situate the reader in context and to introduce the locations, terms, and concepts that are discussed throughout the book. The main subjects addressed are geography, geology, climate, hydrology, soils, land use, flora, vegetation, archaeological heritage, and conservation of natural systems and cultural legacy. Emphasis is placed on the different ecological and landscape features before and after human settlement—albeit not in the processes and causes involved, which are discussed in further chapters—and the handicap that this deep transformation may represent for paleoecological reconstruction. The second chapter summarizes the main prehistorical developments of the ancient Rapanui society. The term "prehistory" has been adopted in Easter Island to refer to the time between human settlement and European contact (Mulloy, 1974). This chapter deals mainly with human settlement and the cultural developments of the ancient Rapanui society. Relevant topics are: the time of arrival and the origin of the first settlers, some hints on sociopolitical organization and demography, the timing of deforestation, the cultural revolution that represented the shift from the moai cult to the Birdman cult, and the debate between ecocidal and genocidal theories. Chapter 3 is an introduction to the study of Easter Island's paleoecology that provides the elements necessary to understand the inception of this discipline in the study of the island's prehistory and considers all paleoecological archives available and all studies carried out in them to date. This chapter subdivides the paleoecological study of Easter Island into three main stages, namely the pioneering phase (1977–1992), the transitional phase (1993–2004), and the revival phase (2005–2019). This provides the basis for the next three chapters, which are organized historically. In Chapter 4, the pioneer work of John Flenley and his colleagues is explained in some detail, with emphasis on deforestation and its possible incidence on cultural change. At the end of the chapter, some insights are provided on points that remained unsolved or were controversial and fostered further paleoecological research. The transitional phase, which represented a paleoecological impasse, is addressed in Chapter 5. During this phase, paleoecological study did not progress significantly due to the lack of systematic fieldwork (mainly sediment coring) and the few papers published. The revival phase (Chapter 6) began with an intense coring effort, between 2005 and 2009, with the participation of several research teams, followed by a substantial increase in publication several years later (2012–2019). All previous coring sites were revisited in a more systematic manner and novel relevant information was obtained on deforestation and land use patterns, useful to complement archaeological and ethnological research.

Chapter 7 summarizes the most relevant paleoecological knowledge obtained during the three phases mentioned and its relevance for a better understanding of Easter Island's prehistory, especially in aspects such as settlement, climate change, deforestation, and cultural change. Chapter 8 presents a conceptual system called EHLFS (Environmental-Human-Landscape Feedbacks and Synergies) able to analyze Easter Island's prehistory from a holistic perspective. The EHLFS system is first explained in some detail and is then applied to Easter Island using the paleoecological evidence summarized in Chapter 7. The epilogue is a call for holistic and truly multidisciplinary study of Easter Island, as the only way of attaining the old aspiration of understanding its prehistorical developments.

Appendix: Supplementary material

The following is the Supplementary material related to this chapter: Please see the video Rapanui 2008 at https://youtu.be/zTbW4kb-DOU.

Rapanui 2008: Images from different sites and monuments of Easter Island taken during the 2008 coring campaign. Photos and composition: V. Rull. Music: Vangelis.

CHAPTER 1

The island at present

Contents

1.1	Geography and geology	1
1.2	Climate and hydrology	5
1.3	Soils and land use	11
1.4	Flora and vegetation	12
	1.4.1 Lichens and bryophytes	12
	1.4.2 Vascular flora	14
	1.4.3 Vegetation	19
	1.4.4 Paleoecological implications of landscape degradation	25
1.5	Archaeological heritage: An outdoor museum	27
1.6	Conservation: The Rapa Nui National Park	33
	1.6.1 Conservation of archaeological heritage	36
	1.6.2 Conservation of native and Polynesian-introduced flora	39

This chapter summarizes the present state of the island in terms of geography, geology, climate, hydrology, soils, flora, vegetation, archaeological heritage, and conservation status. Emphasis is placed on terrestrial landscape and ecosystems, especially flora and vegetation, as the basis for a better understanding of further paleoecological discussions, which are mainly based on pollen analysis. Given the present anthropogenic degradation of the island's biota and ecosystems, a section is dedicated to discussing the potential handicaps imposed by this landscape deterioration for paleoecological reconstruction.

1.1 Geography and geology

Easter Island is located in the southeastern Pacific Ocean, between about 27°03′17″–27°11′42″S and 109°27′00″–109°13′40″W. Geographically, the island is a part of Polynesia, but politically it belongs to Chile (South America) since 1888. Administratively, the island is a special territory of the Chilean Valparaíso region with a certain degree of political autonomy. According to the 2017 census, the island has more than 7700 residents or permanent inhabitants (excluding tourists), most of them living in the

Paleoecological Research on Easter Island
https://doi.org/10.1016/B978-0-12-822727-5.00001-5

capital, Hanga Roa (Fig.1.1). About 45% of these inhabitants consider themselves as Rapanui, that is, descendants of the aboriginal Polynesian colonizers, whereas the others are mainly of Chilean origin, although genetic mixing between these population groups is common (McLaughlin, 2007). It is estimated that approximately 80,000 tourists visit the island every year (Boersema, 2015a).

According to the local tradition, the first name of the island, given by the first Polynesian settlers, was Te Pito O Te Henua, meaning "The Navel of the World" or "The End of the World," depending on the translator. The name "Easter Island" is the English translation of the Dutch name "Paasch Eyland," coined by the first Europeans to arrive in the island, a Dutch expedition led by Jacob Roggeveen who landed on April 5, 1722 (Easter Day). This name is the most popular one and each language uses its

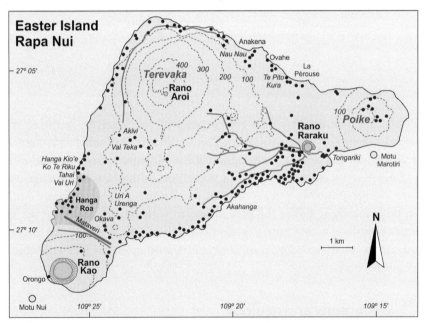

Fig. 1.1 Topographic sketch map of Easter Island. *Dotted lines* represent elevations at 100-m intervals. *Blue circles* indicate the permanent freshwater bodies (Rano Aroi, Rano Kao, and Rano Raraku). The approximate extension of the capital (Hanga Roa) is represented in *gray*. The straight *gray line* is the Mataveri international airport. *Red dots* indicate the approximate distribution of present-day moai and ahu, combining maps from Van Tilburg (1994) and Lipo et al. (2013). The numerous moai still remaining at the Rano Raraku quarry (Fig. 2.10) cannot be represented at this scale. *Green lines* are part of the road network used by the ancient Rapanui for moai transportation, as reconstructed by Lipo and Hunt (2005).

own translation (e.g., "Isla de Pascua" in Spanish, "Île de Pâques" in French, "Isola di Pasqua" in Italian, "Ilha de Páscoa" in Portuguese, or "Osterinsel" in German). The name "Rapa Nui" appeared much later, in the 1860s, and was coined by the Tahitian sailors who came from the Society Islands (Fig. 1 of the Introduction). These sailors knew the very small island of Rapa (also Rapa Iti or "Small Rapa") and referred to Easter Island as Rapa Nui, meaning "Big Rapa." This name was adopted by the aboriginal Easter Islanders, who refer to themselves, their ancestors, and their cultural features as Rapanui.

The tiny Easter Island (164 km^2) has a triangular shape due to the coalescence of three major volcanoes: the Kao, the Poike, and the Terevaka, the latter being the highest summit of the island (511 m elevation), followed by the Poike (370 m) and the Kao (324 m) (Fig. 1.1). In addition, about 100 minor cones are widespread over the island, especially in the southern flank of the Terevaka volcano (Vezzoli and Acocella, 2009). The SE coasts are flat and irregular and spiked by small bays, whereas the other coastal sectors are characterized by high and steep cliffs due to marine erosion, especially in the Kao and Poike surroundings, where these escarpments attain a height of 100–300 m. The only two sandy beaches of the island, Anakena and Ovahe, are located in the NE sector (Fig. 1.1). Easter Island is the emerged part of a large volcanic complex rising from the seafloor at more than 2000 m depth, as part of the Easter Seamount Chain (ESC), a 2500-km W-E alignment of volcanic seamounts situated in the westernmost part of the Nazca Plate (Fig. 1.2). The other emerged element is the small Salas y Gómez Island, an islet of 0.15 km^2 surface and 30 m of maximum elevation, situated at approximately 390 km E-NE of Easter Island (Fig. 1.2). The ESC has been originated by hotspot volcanism, with the Easter hotspot resulting from the activity of the Easter microplate (Haase and Devey, 1996), and the seamounts are older toward the east due to the W-E movement of the Nazca Plate from the spreading East Pacific Rise (EPR). This spreading occurs at a very fast rate of approximately 15 cm/year (DeMets et al., 1994), as the oldest members, situated eastward, date back to the Early Miocene (22–23 million years before the present or Ma) (Smith, 2003; Ray et al., 2012), whereas the youngest submarine seamounts, located between Easter Island and the Easter hotspot, are of the Pleistocene age (0.6–0.2 Ma) (O'Connor et al., 1995). Easter Island is also among the younger members of the ESC, with an oldest age of 0.8 Ma.

The three major volcanoes of Easter Island (Kao, Poike, and Terevaka) experienced a similar and nearly coeval evolution that can be summarized

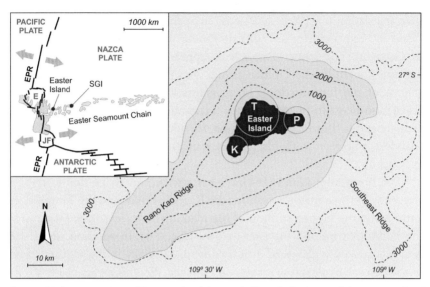

Fig. 1.2 Tectonic setting of Easter Island (upper left) and close-up of the Easter Island seamount with the submerged part in *yellow* and the emerged part (Easter Island) in *red*. Isobaths are represented by dotted lines in 1000 m intervals. The emerged part of the volcanic complex is shaped by the fusion of three volcanoes: the Kao (K), the Poike (P), and the Terevaka (T). *Gray circles* indicate the young fields of volcanic activity. Abbreviations in the tectonic setting (upper left): E, Easter microplate; EPR, East Pacific Rise; JF, Juan Fernández microplate; SGI, Salas y Gómez Island. *(Redrawn and modified from Vezzoli, L., Acocella, V., 2009. Easter Island, SE Pacific: and end-member type of hotspot volcanism. Geol. Soc. Am. Bull. 121, 869–886.)*

into three main phases: (1) buildup of a basaltic shield (0.8–0.4 Ma), (2) formation of a summit caldera (0.4–0.3 Ma), and (3) rifting along the shield flanks by means of fissure eruptions (0.3–0.1 Ma), which formed the numerous minor cones mentioned before (Vezzoli and Acocella, 2009) (Fig. 1.3). The Kao is the only one of the three main volcanoes that conserves a clear summit caldera (Fig. 1.6), with steep inner walls 200 m high and 50–60 degrees slope. The Terevaka is, by far, the largest volcanic edifice and the main contributor to the island's surface (Fig. 1.3). This volcano is also the one responsible for most of the minor cones of the island, including the Aroi cone. The Raraku cone (Fig. 1.1) deserves special attention because it is one of the two tuff (volcanic ash) cones of the island and was the quarry of almost all the moai (Section 1.5). This crater was derived from the activity of the Poike volcano and its minimum age has been estimated at 0.2 Ma (Vezzoli and Acocella, 2009). This volcanic history was responsible for the formation of a variety of volcanic rocks of diverse physicochemical characteristics

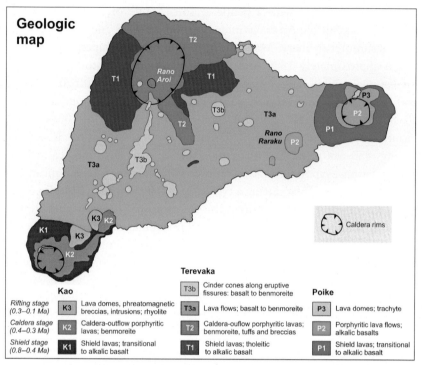

Fig. 1.3 Geologic map of Easter Island. The three volcanoes and the rocks formed by their activity are represented in different colors: the Kao (K) in *red*, the Terevaka (T) in *green*, and the Poike (P) in *blue*. Numbers indicate the stage of volcanic activity, namely the shield stage (1), the caldera stage (2), and the rifting stage (3). Note that the major part of the island is covered by rocks that originated in the Terevaka volcano (T1–T3). The position of Rano Aroi and Rano Raraku is also indicated. *(Redrawn and modified from Vezzoli, L., Acocella, V., 2009. Easter Island, SE Pacific: and end-member type of hotspot volcanism. Geol. Soc. Am. Bull. 121, 869–886.)*

(Fig. 1.3) that, in the absence of metals, constituted a key resource for the development of the Rapanui culture. Baker (1998) and Gioncada et al. (2010) provided a detailed account of the uses of Easter Island's rocks by the ancient islanders, whereas Beardsley and Goleš (1998), Ayres et al. (1998), and Simpson and Dussubieux (2018) focused on the main quarries for the different rock types.

1.2 Climate and hydrology

The present–day climate on Easter Island is subtropical, with small seasonal temperature variations due to the oceanic influence. The annual average temperature is 21°C with a gentle seasonal range of 5–6°C on average, between 18°C in the Austral winter (July–September) and 23–24°C in the Austral summer (January–March) (Fig. 1.4). Extreme temperatures vary between approximately 15°C and 28°C (Herrera and Custodio, 2008). The total annual rainfall ranges between 1100 and 1300 mm, with an average of 140 rainy days. The average seasonal variability ranges between minima of 70–80 mm/month (November–December) and maxima of 100–130 mm/month (April–June) (Fig. 1.4). Potential evapotranspiration is approximately 850–950 mm/year; therefore, the climatic hydrological balance (precipitation-evapotranspiration ratio or PE) is above 1, which means that there is no water deficit during a typical year (Herrera and Custodio, 2008).

A recent survey using short-term observations (<2 years) identified a linear lapse rate in both temperature and precipitation values across the island, from coastal localities to the Terevaka uplands (Puleston et al., 2017). According to these estimates, annual average temperatures were approximately 21°C on coastal sites, with a decrease of -0.85°C per 100 m elevation ($r^2 = 0.99$), whereas precipitation increased at a rate of 175 mm per 100 m elevation from coastal values of 1240 mm ($r^2 = 0.87$). Based on a

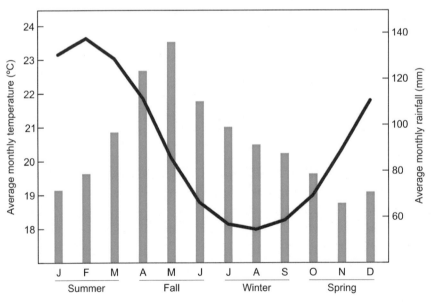

Fig. 1.4 Climatic diagram of Easter Island for the period 1950–2000. Average monthly temperature is represented by a *red line* and average monthly rainfall by *blue bars*. *(Redrawn and modified from Caviedes, C.N., Waylen, P.R., 2011. Rapa Nui: a climatically constrained island? Rapa Nui J. 25, 7–23.)*

detailed elevation model of the island and the predominant E-W direction of trade winds, the same study proposed the existence of a possible rain shadow effect that disturbed the orographic precipitation pattern. This would affect the leeward sides of the main mountains, such as Poike and Terevaka (and, to a lesser extent, of the Kao crater), where total annual precipitation could be <850 mm/year. This value is in the range of the potential annual evapotranspiration values mentioned before; hence, these areas might be close to the boundary of the water deficit.

The seasonal variability in precipitation is controlled by the interplay between three climate systems: the South Pacific Anticyclone (SPA), the South Pacific Convergence Zone (SPCZ), and the westerly storm tracks (Fig. 1.5). During the Austral fall/winter, the weakening and northern migration of the dry SPA favors the progress of humid westerlies, which carry abundant moisture after their passage by the SPCZ, toward the island, thus causing the April–June precipitation increase. During the Austral spring/summer, the SPA migrates southward and prevents the westerly storm fronts from reaching Easter Island, which receives minimum precipitation values (Garreaud and Aceituno, 2001; Sáez et al., 2009). Long-term interannual

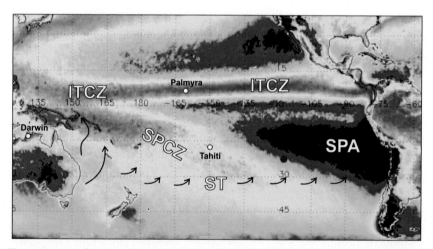

Fig. 1.5 Main climatic systems of the South Pacific (ITCZ, Intertropical Convergence Zone; SPCZ, South Pacific Convergence Zone; SPA, South Pacific Anticyclone; ST, Storm Tracks). Colors indicate precipitation gradients from <100 *(black)* to >1200 *(red)* mm/year for the period 1987–2003. Easter Island is represented as a *red dot*. *(Composed and modified from Negri, A.J., Adler, R.F., Shepherd, J.M., Huffman G., Manyin, M., Neklin, E.J., 2004. A 16-year climatology of global rainfall from SSM/I highlighting morning versus evening effects. In: 13th American Meteorological Conference on Satellite Meteorology and Oceanography, Norfolk, P6.16.)*

variability in precipitation is very high, ranging from 500 to more than 1800 mm/year (Azizi and Flenley, 2008). This variability is largely unpredictable, as it has not been possible to associate long-term precipitation changes with interannual periodic climatic systems such as the El Niño Southern Oscillation (ENSO).

According to MacIntyre (2001), who reviewed the ocean-atmospheric dynamics around Easter Island, and Genz and Hunt (2003), who analyzed a local climatic series of 50 years (1950–2000), Easter Island is unaffected by the ENSO cyclicity likely due to its geographical position, where the fluctuations of the Southern Oscillation Index (SOI), a common parameter to estimate the strength of the ENSO activity (Trenberth, 2019), are minimal or near zero. Caviedes and Waylen (2011), using a similar dataset as that of Genz and Hunt (2003), highlighted that, due to its subtropical latitude and its proximity to the SPA (Fig. 1.5), Easter Island is subjected to climatic controls different from those of tropical Pacific islands, where the ENSO is a dominant climatic mode. These authors also argued that the lack of consistent association between ENSO activity and Easter Island's climatic variability may be due in part to methodological constraints. One of these drawbacks is that the SOI is based on the measured sea-level pressure differences between Darwin (Australia) and Tahiti (Fig. 1.5), and Easter Island is situated > 4000 km SE of Tahiti, having different air pressures and climatic controls (Caviedes and Waylen, 2011).

In contrast to precipitation patterns, some records suggest the possibility of an ENSO influence in the sea surface temperature (SST) interannual variability around Easter Island. For example, a study of SST series around the island for the period 1982–1999 showed that negative SST anomalies coincided with ENSO events (Glynn et al., 2003). Additional support was obtained using paleoecological methods to estimate modern SSTs. An annually banded coral core obtained in the Ovahe beach (Fig. 1.1) provided an oxygen isotope record (a proxy for SST) for the period 1944–1997. The spectral analysis of this record showed that the main source of SST variability was the annual cycle (Mucciarone and Dunbar, 2003). However, a small peak around the 4-year band, which is within the ENSO frequency (Trenberth, 2019), suggested the potential influence of this climatic mode on interannual SST variability.

Despite the humid nature of Easter Island's climate, surface freshwater sources are scanty. Permanent surficial water currents and springs are absent due to the high permeability of the island's volcanic rocks and the existence of abundant fractures (Herrera and Custodio, 2008). After a rain event,

Raraku are fed solely by precipitation and are hydrologically disconnected from the groundwater system. Rano Aroi, situated at 430 m elevation, is an exception, and its swamp is thought to be fed by groundwater penetrating the volcanic core (Herrera and Custodio, 2008). The three ranos are also the only sites with aquatic sediments suitable for paleoecological investigations. The main characteristics of these sediments and their use for paleoecological research are described in more detail in Chapter 3.

1.3 Soils and land use

Soils are developed by weathering of lavas and volcanic ashes and are usually shallow (< 45 cm depth) and covered by loose volcanic pebble accumulations. Soil depth decreases with slope, being maximum in the coastal lowlands (slopes of 1%–3%) and minimum or inexistent with slopes of 40% or more, which are frequent within volcanic cones. Easter Island's soils are composed mainly of silt and clay, which represent about 80%, in average, of the mineral fraction (Alcayaga and Narbona, 1969). Due to the high soil permeability and the rainy climate, lixiviation is high and therefore base saturation is low (30%–35%), with Mg as the most abundant cation due to the influence of marine sprays. As a consequence, soils are acidic and conductivity is low (Table 1.1). K and soluble P are generally low and organic

Table 1.1 Selected physicochemical parameters measured in 42 soil samples from different environments widespread across Easter Island.

Parameter	Range
Slope (%)	1–30
Sand (%)	4–48
Silt (%)	19–42
Clay (%)	20–72
Apparent density (g/cm^3)	0.7–1.2
Field capacity (%)	31.3–55.1
pH	5–6.6
Conductivity (dS/m)	0.08–0.69
N (mg/kg)	21–67
P (mg/kg)	1–355
K (mg/kg)	10–1008
Organic matter (%)	1.3–21.6
Cation exchange capacity (meq/100 g^1)	23.1–55.2

Raw data from Flores, J.P., Torres, C., Martínez, E., Muñoz, P., 2013. Determinación de la Erosión Actual y Potencial de los Suelos en la Isla de Pascua. Ministerio de Agricultura, Centro de Información de Recursos Naturales, Santiago de Chile.

matter may be abundant, especially in the upper horizon (CONAF, 1997). Table 1.1 shows some selected physicochemical parameters measured in soils distributed across the island. Due to the long-standing degradation of vegetation (Section 1.4.3), soil erosion by rain and wind is relatively high and, in some sectors, such as the Poike volcano, may constitute a serious problem for plant growth and agriculture. At present, the surface of eroded soils is approximately $110\,km^2$, which represents almost 67% of the island's surface (Flores et al., 2013).

Of the total surface of the island's soils, it has been considered that about 30% ($48\,km^2$) may be useful for cultivation, 32% ($53\,km^2$) are better suited for pastures, and 38% ($63\,km^2$) are more adequate for forestry and wildlife (Flores et al., 2013). However, only a small proportion of the soils considered suitable for cultivation are actually dedicated to this activity. Cultivation is practiced in small private urban and rural plots totalizing about $7\,km^2$, which is about 4% of the total island's surface (Fig. 1.9) and $<15\%$ of the soil surface considered suitable for agriculture. The main products are vegetables, corn, banana, pineapple, sugarcane, and fruit trees. Local production is insufficient to supply the internal demand of the island, which is compensated from continental Chile. The situation is better for animal husbandry, mainly cattle and horses, as about 50% of the suitable soil surface is used for this purpose. The main economic support of the island is tourism, which is largely based on the prehistoric cultural heritage and outcompetes any other commercial activity (CONAF, 1997).

1.4 Flora and vegetation

The better known plants of Easter Island are the angiosperms and virtually all species of this group have been cataloged. Gymnosperms are very scarce and are usually cultivated. Pteridophytes, which used to be restricted to the ranos and other humid environments, are relatively well known but still need more research. Nonvascular plants (lichens, hornworts, and mosses), on the other hand, are poorly known due to insufficient sampling. A summary of these major groups is provided in the following.

1.4.1 Lichens and bryophytes

Follmann (1961) mentioned some lichens growing on the surface of the moai such as *Lecidea paschalis* (Lecideaceae), *Diploschistes anactinus* (Thelotremataceae), and *Physcia ahu* (Physciaceae) (see also Section 1.6.1). Some of these species have been used to estimate the lichenometric ages of some moai, giving results of 380–460 years (1490–1570 CE), with errors between 7% and 8% (Rutherford et al., 2008). The most complete record of lichens for the island was published by Elix and McCarthy (1998) and contains about 50 species.

The bryophytes were studied by Ireland and Bellolio (2002), who reviewed the former literature and performed an extensive field sampling across the island to report the occurrence of one hornwort (Anthocerotophyta), 11 liverwort (Marchantiophyta), and 30 moss (Bryophyta) species. Hornworts and liverworts from these collections were studied by Grolle (2002). The hornwort remained unidentified and was reported as "Anthocerotaceae spec.," whereas all liverworts but one were identified at species level (Table 1.2). None of the reported taxa were endemic to Easter Island and all of them were widely distributed. Regarding habitat features, the preferred sites were earth, rock, and bark (Grolle, 2002). No mention was made by this author to the native or introduced character of liverwort species.

The 30 moss species reported by Ireland and Bellolio (2002) belong to 24 genera and 19 families (Table 1.3). Only three of these species are endemic to the island. The authors did not mention which species are native to the island but pointed out that 50% of species were shared with continental Chile and 43% with Hawai'i (Table 1.3). The environmental conditions required by these species were not specified either but, since they are mosses, they are supposed to have been collected in moist microenvironments.

Table 1.2 Liverworts from Easter Island identified by Grolle (2002) after collections by Ireland and Bellolio (2002).

Family	Taxa	Distribution
Acrobolbaceae	*Marsupidium knightii*	Asiatic-Pacific-Australasian
Aneuraceae	*Riccardia tenerrima*	Patagonian
Cephalloziellaceae	*Cephaloziella* sp.	NA
Frullaniaceae	*Frullania ericoides*	Pantropical-subtropical
Jackiellaceae	*Jackiella javanica*	Asiatic-Pacific-Australasian
Lejeuneaceae	*Cololejeunea minutissima* ssp. *myriocarpa*	Pantropical-subtropical
	Lejeunea flava	Pantropical-subtropical
	Lejeunea minutiloba	Neotropical
Lophocoleaceae	*Lophocolea aberrans*	Neotropical
Marchantiaceae	*Dumortiera hirsuta*	Pantropical-subtropical
	Marchantia berteroana	Eury-circumsubantarctic

NA, not ascertained.

Table 1.3 Moss species from Easter Island reported by Ireland and Bellolio (2002).

Family	Species	Chile	Hawai'i
Bartramiaceae	*Philonotis hastata*		×
Bryaceae	*Brachymenium indicum*	×	×
	Bryum argenteum		×
	Bryum sp.		
Bruchiaceae	*Trematodon pascuanus*[a]		
Dicranaceae	*Dicranella cardotii*	×	
	Dicranella hawaiica		×
	Anisothecium hookeri	×	
Ditrichaceae	*Ceratodon purpureus*	×	×
	Ditrichum dificile	×	
Fabroniaceae	*Fabronia jamesonii*	×	
Fissidentaceae	*Fissidens pascuanus*[a]		
	Fissidens pellucidus		
Hypnaceae	*Isopterygium albescens*		×
Leucobryaceae	*Campylopus clavatus*	×	
	Campylopus introflexus	×	×
	Campylopus vesticaulis	×	
Meesiaceae	*Leptobryum pyriforme*	×	×
Meteoriaceae	*Papillaria crocea*		
Mniaceae	*Pohlia* sp.		
Orthotrichaceae	*Macromitrium* sp.		
Pottiaceae	*Chenia leptophyla*		×
	Tortella humilis		×
	Weissia controversa	×	×
Ptychomitriaceae	*Ptychomitrium subcylindricum*[a]		
Racopilaceae	*Racopilum cuspidigerum*	×	×
Rhizogoniaceae	*Pyrrhobryum spiniforme*		×
Seligeriaceae	*Blindia magellanica*	×	
Sematophyllaceae	*Sematophyllum aberrans*	×	
	Sematophyllum brachycladulum	×	

[a] Endemic species.

1.4.2 Vascular flora

According to a recent study, there are 16 pteridophyte species/varieties native to Easter Island, of which 4 species and 1 variety are endemics (Meyer, 2013). Four of these species (one endemic) might be considered extinct, as they have not been observed or collected since more than 75 years ago. All these species are typical of moist sites, especially within and around crater lakes and swamps (Table 1.4). In addition, four other species are reported by the same author as introduced and cultivated in the gardens of Hanga Roa

Table 1.4 Pteridophyte species native to Easter Island and some of their main features.

Family	Species	Distribution	Habitat (EI)
Aspleniaceae	*Asplenium polyodon* var. *squamulosum*[a]	Asia, Australia, Polynesia (species)	Peat moss (Rano Kao)
	Asplenium obtusatum var. *obtusatum*	South America, Australia, New Zealand	Rocks near the sea shore, cliffs (widespread)
Athryriaceae	*Diplazium fuenzalidae*[a]	EI	NA
Blechnaceae	*Blechnum paschale*[a]	EI	Humid rock fissures and crevices
Davalliaceae	*Davallia solida*[b]	Pantropical	Rano Kao
Dennstaedtiaceae	*Microlepia strigosa*	Asia, Polynesia	Widespread
Dryopteridaceae	*Dryopteris karwinskyana*[b]	Tropical America	Among stones and boulders (Rano Aroi)
	Elaphoglossum skottsbergii[a]	EI	Among stones and boulders, rock fissures (Rano Aroi)
	Polystichum fuentesii[a,b]	EI	NA
Ophioglossaceae	*Ophiglossum lusitanicum* subsp. *coriaceum*	Australia, New Zealand, South America	Moist stones under dense grass cover
	Ophioglossum reticulatum	Pantropical	Moist stones under dense grass cover
Polypodiaceae	*Microsorum parksii*	Pacific	Rock fissures (Rano Kao, Rano Aroi)
Psilotaceae	*Psilotum nudum*[b]	Pantropical	Rano Aroi?
Pteridaceae	*Haplopteris ensiformis*	Asia, Australia, Polynesia	Among stones and boulders, rock fissures (Rano Kao, Rano Aroi)
Thelypteridaceae	*Pneumatopteris costata* var. *hispida*	Polynesia	Among boulders and in caves (Rano Kao, Rano Aroi)
	Thelypteris interrupta	Pantropical	Among boulders and within the reed-swamp zone (Rano Kao, Rano Aroi)

[a] Endemic.
[b] Extinct.
EI, Easter Island; NA, not ascertained.
Raw data from Meyer, J.Y., 2013. A note on the taxonomy, ecology, distribution and conservation status of the ferns (Pteridophytes) of Rapa Nui (Easter Island). Rapa Nui J. 27, 71–83.

and Orongo. These are: *Adianthum* cf. *raddianum* (Adianthaceae), *Cyrtomium falcatum* (Dryopteridaceae), *Nephrolepis* cf. *cordifolia* (Nephrolepidaceae), and *Cyclosorus* cf. *parasiticus* (Thelypteridaceae). There are no gymnosperms in the wild; only some species of *Araucaria* (Araucariaceae), *Cupressus* (Cupressaceae), and *Cedrus*, *Larix*, and *Pinus* (Pinaceae) are cultivated around the capital, Hanga Roa (Zizka, 1991).

According to Zizka (1991), the known angiosperm flora of the island is composed of about 180 species, of which <17% are considered to be autochthonous (idiochores) (Table 1.5) and almost 80% have been introduced by humans (anthropochores), the remaining 4% being of uncertain origin. Given the isolation of Easter Island, the ways in which the idiochore species reached the island are of special significance. Interestingly, none of the autochthonous angiosperms has potential for wind dispersal, the main potential mechanisms being transportation by birds (75%) and drift through seawater (25%) (Carlquist, 1967, 1974). Among the introduced species, almost half are considered to be naturalized and are permanent elements of the wild vegetation (agriophytes), whereas the other half occur temporarily on anthropogenic sites and still depend on human activities for their subsistence (ephemerophytes). Most of the anthropochore species have been introduced after European contact, and only a few species were carried by the first Polynesian settlers for cultivation purposes (Section 1.4.3). Globally, the families with more species are Poaceae (43 species), Asteraceae (18), Fabaceae (17), and Brassicaceae (7), which are also the families with more introduced species. The families with more autochthonous species are Poaceae (9 species) and Cyperaceae (5), followed by Convolvulaceae, Fabaceae, and Solanaceae, with two species each.

Only 3(4) extant species have been considered endemic to the island: *Axonopus paschalis* (Poaceae), *Danthonia paschalis* (Poaceae), *Sophora toromiro* (Fabaceae), and probably *Paspalum forsterianum* (Poaceae) (Zizka, 1991). Among them, the toromiro (*S. toromiro*) (Fig. 1.8) is considered to be extinct in its natural habitat for the past 50 years and is only maintained by cultivation in the island (Fig. 2.18) and in several botanical gardens elsewhere (Maunder et al., 2000). A recent review of Poaceae by Finot et al. (2015) reported 51 grass species, of which 8–9 were most probably native and at least 42 were introduced. According to these authors, only two of the native species, *Rytidosperma paschale (= Danthonia paschalis)* and *Paspalum forsteianum*, were endemic to the island. *Axonopus paschalis*, which was previously considered as endemic to Easter Island (Table 1.3), was transferred by Finot et al. (2015) to *Axonopus compressus*, native to South America, thus losing its endemic character.

Table 1.5 Main features of the angiosperm species considered to be autochthonous (idiochores) in Easter Island (Zizka, 1991).

Family	Species	Common name (EI)	Distribution	Habit	Habitat (EI)
Aizoaceae	*Tetragonia tetragonoides*	Herepo	Widespread	Annual herb	Coastal sites and islets
Apiaceae	*Apium prostratum*	NA	Subantarctic	Perennial herb	Coastline (rare)
Arecaceae	*Paschalococos disperta*[a,b]	NA	EI	Palm tree	Widespread
Chenopodiaceae	*Chenopodium glaucum*	NA	Widespread	Annual herb	Sea shore
Convolvulaceae	*Calystegia sepium* (?)	NA	Widespread	Perennial herb	Rano Kao and Rano Aroi
	Ipomoea pes-caprae	Tanoa	Widespread	Perennial herb/shrub	Sea shores (rare)
Cyperaceae	*Cyperus cyperoides*	Hikukio'e	Widespread	Perennial herb	Wet places
	Cyperus eragrostis	Hikukio'e	American	Perennial herb	Ranos and other wet places
	Kyllinga brevifolia	Hikukio'e	Widespread	Perennial herb	Higher ground and wet places
	Pycreus polystachyos	Hikukio'e	Widespread	Annual or short-lived perennial herb	Damp places and higher ground
	Scirpus californicus	Ngaatu	American	Perennial herb	Crater lakes
Euphorbiaceae	*Euphorbia serpens*	Pato	American	Annual herb	Overgrazed or disturbed ground
Fabaceae	*Caesalpinia major*	Naoho, Ngaoho	Widespread	Shrub	Rano Kao (rare)
	Sophora toromiro[a,b]	Toromiro	EI	Shrub or small tree	NA
Gentianaceae	*Centaurium spicatum* (?)	NA	Widespread	Annual or short-lived perennial herb	Near the coast
Poaceae	*Agrostis avenacea*	Tuere heu	SE Asia–Australia–Pacific	Annual or short-lived perennial herb	Common
	Axonopus paschalis[a]	Heriki hare	EI	Perennial herb	Higher elevations (Terevaka)

Continued

Table 1.5 Main features of the angiosperm species considered to be autochthonous (idiochores) in Easter Island (Zizka, 1991)—cont'd

Family	Species	Common name (EI)	Distribution	Habit	Habitat (EI)
	Bromus catharticus	NA	American	Annual or short-lived perennial herb	Cultivated ground and along waysides
	Cynodon dactylon (?)	Mati	Widespread	Perennial herb	Poike
	Danthonia paschalis[a]	NA	EI	Perennial herb	Rano Kao
	Dichelachne crinita	NA	SE Asia–Australia–Pacific	Perennial herb	Rare
	Dichelachne micrantha	NA	Australia–Pacific	Perennial herb	NA
	Paspalum forsterianum[a] (?)	Haiki hare	EI	Perennial herb	Coastal cliffs and islets
	Sporobolus africanus	Here hoi	Widespread	Perennial herb	Frequent. Grazed areas in the S and E
Polygonaceae	*Polygonum acuminatum*	Tavari	American	Perennial herb	Margins of the crater lakes
Portulacaceae	*Portulaca oleracea*	NA	Widespread	Annual herb	Sea shores, waysides and waste places
Primulaceae	*Samolus repens*	NA	Subantarctic	Perennial herb	Rare
Solanaceae	*Lycium carolinianum*	Pua nako nako	Pacific	Shrub	Rare
	Solanum forsteri	Poporo	Pacific	Herb or small shrub	Rare
Tiliaceae	*Triumfetta semitriloba*	Hau hau	American	Shrub or small tree	Rare

[a] Endemics.
[b] Extinct.
EI, Easter Island; NA, not ascertained.

Fig. 1.8 Toromiro (*Sophora toromiro*). *(Reproduced under license CC BY-SA 4.0. Image source: https://fi.wikipedia.org/wiki/Toromirohelmipalko.)*

Considering pteridophytes and angiosperms together, the native vascular flora of Easter Island is composed of 45 species/varieties, 10 of which are endemics and 6 (2 of them endemics) may be extinct (Zizka, 1991). Dubois et al. (2013) reported similar figures with 48 native species/varieties, 11 endemics and 21 probably extinct. The difference in the number of extinct species is due to the consideration by Dubois et al. (2013) of taxa identified in archaeological and paleoecological remains (charcoal, pollen, spores), which are no longer living on the island (Section 1.4.3). The flora of Easter Island has been considered exceptionally poor, especially in woody species. This was firstly noted by Skottsberg (1920–1956), who observed that the flora of the island was comparatively poorer than that of other oceanic islands of similar size, geology, elevation, and climate, and attributed this phenomenon to the exceptional isolation of Easter Island (see also Dumont, 2002). In contrast, Van Balgooy (1960) believed that many species became extinct as a result of human activities, a hypothesis that was supported later by Flenley et al. (1991), after their pioneering paleoecological studies (Chapter 4).

1.4.3 Vegetation

The vegetation of Easter Island is fully degraded and anthropized. Today, the island is largely covered by grass meadows (90%), with few forest (5%) and shrubland (4%) patches; pioneer and urban vegetation represent 1% of the surface (Etienne et al., 1982). Fig. 1.9 shows more details of these vegetation

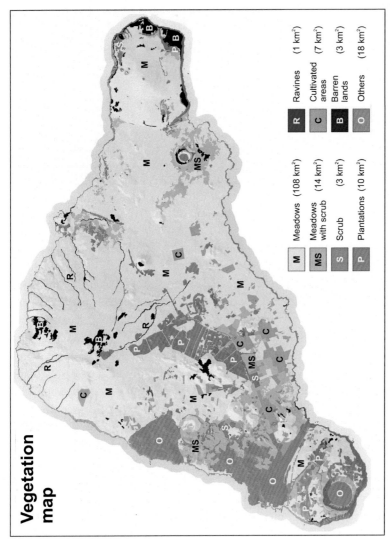

Fig. 1.9 Vegetation map of Easter Island, indicating the approximate total extension of each vegetation type. The category "others" includes urban areas, cliffs, crater walls, and waterbodies, among others. (*Redrawn and modified from CIREN, 2013. Mapa de Vegetación de Isla de Pascua. Centro de Información de Recursos Naturales, Ministerio de Agricultura, Santiago de Chile. Available at https://www.ciren.cl/; Accessed 6 November 2019.*)

units. Meadows have a variable soil cover, ranging from 25% to 50% in the southern coasts to 75%–100% in the N, NW, and E sectors of the island. Most meadows are dominated by two widely distributed grasses, *Sporobolus africanus* (here hoi) and *Paspalum scrobiculatum*, but the Terevaka uplands are dominated by the endemic grass *Axonopus paschalis* (heriki hare) and the sedges *Pycreus polystachyos* and *Kyllinga brevifolia* (hikukio'e). Shrublands, with typical soil covers between 50% and 75%, are largely dominated by the invader *Psidium guajava* or tuava (Myrtaceae), introduced from tropical America. Other relevant shrubs are the also introduced *Crotalaria* spp. (Fabaceae) and *Dodonaea viscosa* (Sapindaceae), probably escaped from cultivation. Forests are all planted and their soil cover is highly variable (50%–100%). Most forests are recent plantations of *Eucalyptus* spp. (Myrtaceae), introduced from Australia (Fig. 1.11), *Dodonaea viscosa*, carried from tropical America, the widely distributed *Melia azedarach* (Meliaceae), and *Thespesia populnea* (Malvaceae), which some authors consider to be an idiochore and others an anthropochore species (Zizka, 1991). Dubois et al. (2013) considered that this species was introduced by Polynesian colonizers. There is also a plantation of coconut (*Cocos nucifera*; Arecaceae) in the bay of Anakena (Fig. 1.11). The vegetation of the ranos is dominated by the aquatic species *Scirpus californicus* (Cyperaceae) and *Polygonum acuminatum* (Polygonaceae), both occurring also in tropical America, along with several native fern species (Table 1.2).

The present vegetation of Easter Island is completely different from the preanthropic vegetation that existed before Polynesian settlement, occurred between 800 and 1200 CE, depending on the authors (Flenley and Bahn, 2003; Hunt and Lipo, 2006; Kirch, 2010; Wilmshurst et al., 2011) (see Chapter 2 for details). Since then, the original landscape has been severely degraded and replaced by a totally cultural landscape. This degradation occurred in two main phases. Before Polynesian arrival, the island was covered by palm-dominated forests that were fully removed before European contact (1722 CE), during the development of the ancient Rapanui culture. The evidence for this deforestation is mostly paleoecological and will be discussed in detail in Chapters 4–6. Polynesians also introduced cultivated plants such as mahute or paper mulberry (*Broussonetia papyrifera*; Moraceae), taro (*Colocasia esculenta*; Araceae), ti (*Cordyline fruticosa*; Asparagaceae), yam (*Dioscorea alata*; Dioscoreaceae), kumara or sweet potato (*Ipomoea batatas*; Convolvulaceae) (Fig. 1.10), ipu kaha or bottle gourd (*Lagenaria siceraria*; Cucurbitaceae) (Fig. 1.10), and maika or banana (*Musa* sp.; Musaceae), as documented by historical reports and fossil remains in soils from former agricultural drylands and lakeshore terraces (Zizka, 1991; Wozniak et al., 2007; Horrocks and Wozniak, 2008;

Fig. 1.10 Examples of emblematic plants cultivated by the ancient Rapanui people. (A) Flowers and tuber of sweet potato (*Ipomoea batatas*), likely one of the main components of the ancient Rapanui diet. (B) Fruit of bottle gourd (*Lagenaria siceraria*), commonly used for water storage and transportation. (From (A) *https://commons.wikimedia.org/wiki/File:Ipomoea_batatas_002.jpg and (B) https://commons.wikimedia.org/wiki/File:Courge_encore_verte.jpg*.)

surface water flows for only hours to a few days (Brosnan et al., 2018). The only stable freshwater bodies are two lakes (Kao and Raraku) and a swamp (Aroi) located within volcanic craters, a combination locally known as rano (Fig. 1.6). There is also a groundwater system where freshwater accumulates on top and salinity increases with depth due to the penetration of seawater from below, which creates a density gradient (Herrera and Custodio, 2008) (Fig. 1.7). This groundwater system is shallower along the coasts, where natural brackish seeps may occur (Brosnan et al., 2018). Therefore, freshwater is more accessible in coastal environments, especially after heavy rainfall events, where wells for human use are common today. Lakes Kao and

Fig. 1.6 Panoramic views of the three permanent freshwater sources of Easter Island, containing sediments suitable for paleoecological research. *(Photos: V. Rull.)*

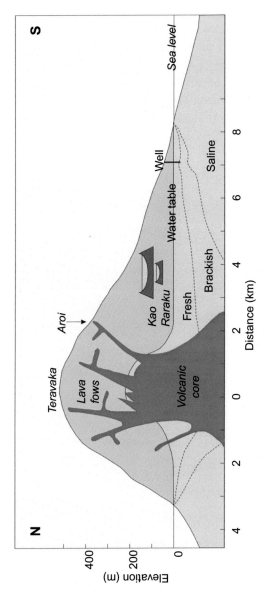

Fig. 1.7 Schematic cross section of an N-S transect showing the hydrological model of Easter Island. Note the progressive thinning of the freshwater table toward the coast. The approximate elevation of lakes Kao and Raraku and the position of the Aroi swamp are indicated. *(Redrawn and simplified from Herrera, C., Custodio, E., 2008. Conceptual hydrogeological model of volcanic Easter Island (Chile) after chemical and isotopic surveys. Hydrogeol. J. 16, 1329–1348.)*

Horrocks et al., 2012a,b, 2013, 2015, 2016). Other cultivated or useful plants mentioned by earlier European visitors were makoi (*Thespesia populnea*; Malvaceae), pua (*Curcuma longa*; Zingiberaceae), toa (*Saccharum officinarum*; Poaceae), and marikuru (*Sapindus saponaria*; Sapindaceae), the last of doubtful origin. Most of these aboriginal introductions, however, are no longer present or cultivated on the island (Zizka, 1991).

Flenley et al. (1991), in their pioneering paleoecological studies, identified several pollen types as belonging to genera likely native to Easter Island or introduced before European contact that are presently extinct on the island. These are: *Acalypha* (Euphorbiaceae), *Coprosma* (Rubiaceae), *Macaranga* (Euphorbiaceae), *Metrosyderos* (Myrtaceae), *Potamogeton* (Potamogetonaceae), *Trema* (Ulmaceae), and *Typha* (Typhaceae). Other significant palynological findings of presently extinct plants are *Dianella* (Xanthorrhoeaceae) in the Rano Raraku sediments (Cañellas-Boltà et al., 2014) and *Sisyrinchium* (Iridaceae) in the Rano Aroi peats and soils (Horrocks et al., 2015). In addition, the pollen of the American weed *Verbena litorais* (Verbenaceae), a species that is now widespread across the island and previously considered to have been introduced after European contact (Zizka, 1991), was found in the sediments of Rano Raraku long before Polynesian arrival (Cañellas-Boltà et al., 2013). The implications of these paleoecological findings for the initial settlement of Easter Island are discussed in more detail in Chapter 7. Charcoal remains from some archaeological sites enabled identification of 17 woody taxa with ages ranging from the early 14th to the mid-17th century (Orliac, 2000; Orliac and Orliac, 1998). Seven of these taxa are part of the present Easter Island's flora, of which four are considered to be native and three to be introduced by Polynesians. The other 10, however, are no longer present in the island and it is not clear whether these fragments corresponded to species from ancient forests (Orliac, 2000), a question that remains open. This will also be further discussed in more detail in Section 5.4, but the final result is that the first Europeans who arrived in Easter Island encountered a totally deforested landscape.

Landscape degradation did not stop with forest clearance and Rapanui land use but was exacerbated after European contact, as clearly documented in historical records (e.g., McCall, 1981; Fischer, 2005; Boersema, 2015a). Initially, several unsuccessful attempts to introduce exotic species such as goats, pigs, sheep, corn, and a number of fruit trees were carried out. Major deterioration occurred in 1875 CE when the whole island was transformed into a ranch, mostly for sheep. This initiated the second island-wide degradation, after deforestation. Former forest removal had eliminated the palm forests, but intensive and extensive grazing removed most of the autochthonous plant species that remained. After a pause in which more alien species were introduced,

Fig. 1.11 Recent tree plantations on Easter Island. (A) Coconut (*Cocos nucifera*) plot in the bay of Anakena, with the moai of Ahu Nau Nau in the background. (B) Forest stand of *Eucalyptus* on the shore of Rano Aoi. *(Photos courtesy of N. Cañellas.)*

notably coconut trees (*Cocos nucifera*) (Fig. 1.11), a second and even more intense landscape degradation event took place as a result of the reactivation of extensive livestock practices. In 1903, the number of sheep in the ranch—known as "Fundo Vaitea"—increased to approximately 70,000, and the island experienced the worst vegetation deterioration of its entire history.

The introduction of exotic species continued with the planting of *Eucalyptus* forest stands (Fig. 1.11), which are still standing on the island, along with other species for reforestation and ornamental purposes, or to protect soils from erosion. Relevant examples are *Melia azedarach* (Meliaceae), *Robinia pseudoacacia*, *Acacia* spp. and *Crotalaria grahamiana* (Fabaceae), *Grevillea robusta* (Proteaceae), *Hibiscus tiliaceus* (Malvaceae), *Lantana camara* (Verbenaceae), *Catharanthus roseus* (Apocynaceae), *Anredera cordifolia* (Basellaceae), and *Dodonaea viscosa* (Sapindaceae). Other species introduced intentionally were the tobacco, *Nicotiana tabacum* (Solanaceae), for obvious reasons, and *Psidium guajava* (Myrtaceae), for its edible fruits. Some species were introduced to improve pastures, notably the grasses *Melinis* spp., *Pennisetum clandestinum*, and *Setaria sphacelata*. Many other anthropochore species seem to have been introduced unintentionally as weeds. A thorough account of the introduced species is provided by Zizka (1991). Although the island's landscape is entirely anthropized, the areas with less anthropochore species are some inaccessible coastal cliffs, the summit of the Terevaka, the inner slopes of the Kao crater, and the Aroi swamp. The inner Kao slopes deserve special attention, especially in terms of conservation, due to the presence of favorable microclimates and several idiochores such as the endemic *Danthonia paschalis,* the rare *Caesalpinia major* and *Triumfetta semitriloba*, along with the aquatic species *Scirpus californicus* and *Polygonum acuminatum*. It is also noteworthy that the last wild specimens known of *Sophora toromiro* grew there (Zizka, 1991).

1.4.4 Paleoecological implications of landscape degradation

The dramatic landscape deterioration is a handicap for paleoecological research, especially for pollen analysis, for several reasons. It could be expected that the main components of pollen records representing the time between Polynesian settlement and European contact are extant idiochore species, Polynesian introductions and presently extinct species, either autochthonous or introduced by Polynesians. This represents <20% of the present flora, which is the only material available for assembling a pollen reference collection for the identification of pollen types present in the corresponding sediments. Other pollen types locally extinct as a consequence of human activities should be identified by comparison with other floras, provided they occur elsewhere.

A critical case is that of the extinct palm that dominated the ancient forests, whose only remains are pollen, empty fruit endocarps and root molds that could correspond to several palm species. An early detailed morphological analysis suggested that the pollen belonged to a species of the Cocosoidae subfamily, represented in the Pacific region by the widespread *Cocos nucifera*

Fig. 1.12 The Chilean wine palm (*Jubaea chilensis*) growing in the La Campana National Park, in the continental Chilean coasts (Mieth and Bork, 2010). *(Photo courtesy of A. Mieth.)*

(coconut) and *Jubaea chilensis* (Chilean wine palm) (Fig. 1.12), which inhabits the Pacific Chilean coasts (Dransfield et al., 1984). Further discoveries of palm nut endocarps that were significantly smaller than those of coconuts, and of trunk and root casts that were similar to those of *Jubaea* seemed to support the second possibility (Arnold et al., 1990; Grau, 1998, 2001). However, this analogy was questioned based on the conspicuous environmental differences between Easter Island and the Chilean coasts, and a possible affinity with *Juania australis*, an endemic palm of the Juan Fernández Islands, was suggested (Hunter-Anderson, 1998). Based on endocarp morphology, a new monospecific genus was described for Easter Island's palm called *Paschalococos disperta*, which is thought to be extinct (Zizka, 1991). A further phytolith analysis again revealed morphological similarities with *J. chilensis*, but a detailed statistical analysis of the morphometric characters of these phytoliths suggested the possibility that more than one palm species grew on the island (Delhon and Orliac, 2007). The same possibility was suggested by Gossen (2011), who reported several palm pollen types from the Rano Kao sediments.

Another handicap is the difficulty of performing modern-analog studies, which is one of the most common methods for reconstructing past vegetation using pollen analysis (Birks and Birks, 1980). This method is based on the principle of uniformitarianism and assumes that the relationship between pollen assemblages and vegetation is constant over time (on a millennial scale), and hence modern pollen-vegetation relationships may be used to infer past vegetation patterns from past pollen assemblages contained in sediments. There are several techniques of modern-analog studies ranging from qualitative inferences to sophisticated statistical tools able to provide quantitative estimates of past vegetation patterns in terms of abundance and vegetation cover (Sugita, 2007a,b; Birks et al., 2010). The same procedures have been used to estimate past climatic parameters, notably paleotemperatures, from fossil pollen assemblages (Birks et al., 2012). A condition to extrapolate modern pollen-vegetation relationships to the past is that community composition has remained fairly constant through time, but sometimes this is not true and some past communities lack modern analogs due to unexpected environmental shifts or differential (idiosyncratic) responses of individual species to environmental change (Williams and Jackson, 2007). Human activities can also be responsible for the absence of modern analogs for past vegetation due to alien introductions, local extinctions, or partial or total removal/replacement of native vegetation. Easter Island is an extreme case, where preanthropic vegetation has been totally removed and the native flora has been reduced to a minimum. Therefore, modern-analog studies at the community level are unfeasible and the reconstruction of past vegetation relies on inferences based on the known autecology of the plant taxa represented in sedimentary pollen assemblages (Flenley et al., 1991). This approach is only possible for extant taxa and the reliability of inferences depends on the precision of pollen identification, ideally at the species level.

1.5 Archaeological heritage: An outdoor museum

The most unique and impressive feature of Easter Island's landscape is its archeological heritage, a large part of which is openly exposed in situ across the whole island (Fig. 1.13). This is why Mulloy (1974) considered Easter Island as "the most spectacular potential outdoor museum to be found anywhere in Polynesia." Documenting the archaeological record of Easter Island is beyond the scope of this book; only a brief summary of the published information is provided to highlight the magnitude of the still preserved prehistoric heritage, as a manifestation of the amazing cultural development of the ancient Rapanui society. Torres Hochstetter et al. (2011) provided a historical account of the efforts developed to document and catalog the archaeological

Fig. 1.13 Panoramic view of the moai complex of Ahu Tongariki near Rano Raraku, one of the most prominent megalithic monuments of the island, with the Poike cliffs and the Motu Marotiri islet in the background (see Fig. 1.1 for location). These moai are up to 9 m high and > 80 tons of weight. *(Photo: V. Rull.)*

heritage of the island in one single item, an initiative that seems still incomplete. According to these authors, the information is dispersed across a mixed assortment of products including maps, general descriptions, reports, dissertations, papers, and books, but a comprehensive catalog of the archaeological remains of the island is still unavailable. McCoy (1976, 1979) summarized the works initiated in the 1960s and his own field studies into detailed maps. This work was continued by Cristino et al. (1981) who published the first archaeological atlas of Easter Island including additional records, designations, descriptions, photographs, and maps. In a later book, the same authors published the more complete record with detailed descriptions available to date (Vargas et al., 2006). According to these authors, there are more than 20,000 sites or manifestations of archaeological interest on the island, still preserved in their original locations.

These manifestations include the abovementioned moai, of which more than 500 are associated with more than 300 ahu (Fig. 1.13), whereas approximately 400 are still in the Raraku quarry (Fig. 2.10), and about 60 remain isolated along transportation roadways (Lipo and Hunt, 2005; Lipo et al., 2013) (Fig. 1.1). Most of these statues were carved on the comparatively soft tuff of the Raraku crater using carving instruments (toki) made of harder basalt (Gioncada et al., 2010), as the ancient Rapanui did not know metals. Only 13 moai were made of basalt, 22 of trachyte, and 17 of red scoria (Van Tilburg, 1994). A number of moai wear big cylindrical hats (pukao) made of red scoria (Gioncada et al., 2010; Hixon et al., 2018). Some moai and moai fragments have been removed from the island by alien visitors and are preserved in private collections and museums in continental Chile and elsewhere (New Zealand, Great Britain, France, Belgium, the United States, and Canada) (McLaughlin, 2007). Other significant manifestations of the ancient Rapanui culture are the about 4200 remains of dwellings (hare) of the two main types: the hare paenga (Fig. 1.14), consisting of an elliptical base made of basalt to support thatched roofs, and the hare moa, made entirely of basalt. Other stone constructions are the manavai, of which about 2000 have been documented, which were used to protect crops from erosion and evaporation, and the pipi horeko or milestones, to indicate some prohibited sites (tapu) or to divide properties. Special mention should be made of the approximately 4000 petroglyphs (rona) (Fig. 1.15), also known as rock art, with varied motifs including octopuses, sharks, turtles, lizards, chickens, human vulvas, and mythic figures, notably the Birdman, which is especially frequent around Orongo, for reasons that will be explained in the following chapter (Section 2.4.3) (Lee, 1992). According to Charola et al. (1998), Easter Island's rocks art is "…unrivaled on Oceania, and have no peer elsewhere." Other features and objects are earth ovens (umu), portable artifacts

Fig. 1.14 Example of hare paenga near Hanga Roa showing the elliptic stone base and the paved entrance, both made of basalt. *(Photo courtesy of S. Giralt.)*

Fig. 1.15 Petroglyphs (rona) carved on basalt from the ceremonial village of Orongo, in the SW crest of the Kao crater (see Fig. 1.1 for location). *(Photo: V. Rull.)*

such as stemmed obsidian tools (mata'a), lithic flakes and stone adzes, as well as modified caves and small carved basins on bedrock to collect water (taheta) (Hunt and Lipo, 2018). There are also extensive subterranean caves, mostly lava tubes, containing some structural remains, portable artifacts, and human and animal (fish, rat, chicken, bird, shellfish) remains (Rorrer, 1998; Ciszewski et al., 2009). Vargas et al. (2006) also provided a map that, albeit incomplete, shows that the main concentration of archaeological remains is in the coasts, especially in the flatter southern and eastern coasts (Fig. 1.16).

Another source of archaeological material is the Father Sebastian Englert Anthropological Museum (www.museorapanui.gob.cl), with about 15,000 objects including petroglyphs, obsidian flakes (mata'a), tattoo pigments (very common in the Rapanui culture), a diversity of stone tools, necklaces, and other ornaments of shell and bone, talismans, harpoons, fishing hooks, needles, and dresses. Some wood carvings are especially significant, notably the ao, an oar-shaped stick that symbolized the Rapanui authority. Other noteworthy wood carvings are the moai kava kava, small anthropomorphic figures (Fig. 1.17), probably of ritual use.

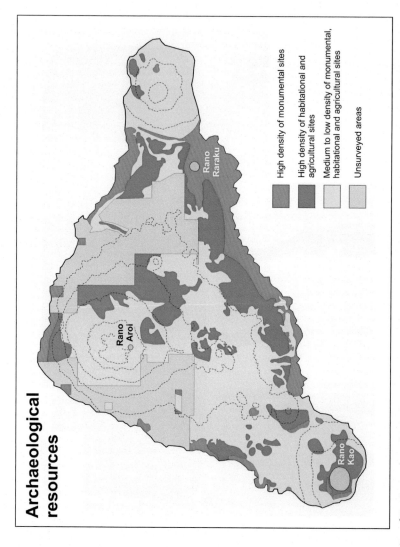

Fig. 1.16 Density of in situ archaeological resources (monumental, habitational, and agricultural features) on Easter Island. The ranos are represented as blue circles. (Redrawn and modified form Vargas, P., Cristino, C., Izaurieta, R., 2006. *1.000 años en Rapa Nui. Cronología del asentamiento. Editorial Universitaria, Santiago de Chile.*)

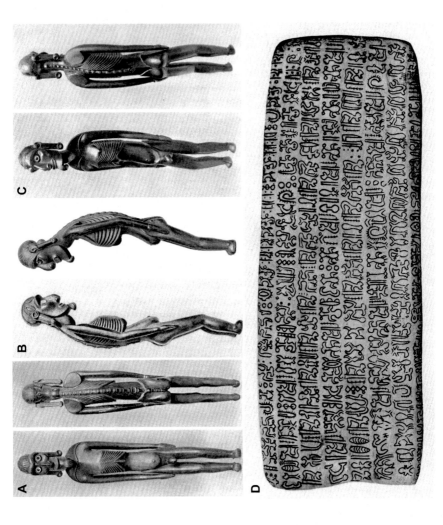

Fig. 1.17 Examples of small Rapanui wood carvings. (A–C) Moai kava kava. (D) Rongorongo tablet. (*Images courtesy of A. Altman and S. MacLaughlin.*)

Several wood tablets with likely symbolic scripts known as rongorongo exist (Fig. 1.16), but both historical studies and radiocarbon dates indicate that this is probably a postcontact feature (Bahn, 1996; Orliac, 2005). The rongorongo script remains undeciphered and many interpretations exist on its possible meaning (e.g., Fischer, 1997; Horley, 2007). The preferred materials for wood carving were the toromiro (*Sophora toromiro*) and the makoi (*Thespesia populnea*) (Orliac, 1990, 2007).

Several online databases exist for Easter Island's archaeological heritage. One is the Rapa Nui Archaeological Database (http://www.rapanuidatabase.org/), which is in construction and contains about 660 moai, 75 pukao, and the roads through which the moai were transported (Torres Hochstetter et al., 2011). This database uses high-resolution Google Earth images to display the content and has been compiled by a team of archaeologists from several American universities (California, Hawai'i, Oregon) and the Father Sebastian Englert Anthropological Museum of Easter Island. The Shepardson Moai Database (http://www.terevaka.net/dc/databases/shepardson_2009/Moai_pt1.html) is also linked to Google Earth for graphical display (Shepardson, 2006, 2009). Combining these two databases, a fairly complete record of more than 962 moai with precise location information (submeter GPS data), photos, and descriptions is available (Schumacher, 2013; Hunt and Lipo, 2018). Shepardson's website also contains a very detailed mata'a database (http://www.terevaka.net/mataa/index.html). Another online database, called the Database of Easter Island Archaeology (https://pages.uoregon.edu/wsayres/NEW/easter_island/UraUranga/uusurveyhome.html), is available at the University of Oregon (United States). The Rapa Nui Interactive Radiocarbon Database (http://data.bishopmuseum.org/C14/) of the Bishop Museum in Hawai'i contains the available radiocarbon ages on archaeological materials from the island (Mulrooney, 2013). The Easter Island Statue Project Website (http://www.eisp.org/category/archaeology/moai-inventory/) is a comprehensive database of moai data collected by Van Tilburg (1994). However, the goal of developing a comprehensive database of the full archaeological record spread over the island, either in print or online, seems still to be pending (Torres Hochstetter et al., 2011).

1.6 Conservation: The Rapa Nui National Park

To protect Easter Island's cultural and natural heritage, the Rapa Nui National Park was created in 1935, although its present boundaries were established in 1995. The same year, the United Nations Educational, Scientific, and

Cultural Organization (UNESCO) declared the park as a World Heritage Site, based on the following criteria (https://whc.unesco.org/en/list/715):

Rapa Nui National Park contains one of the most remarkable cultural phenomena in the world. An artistic and architectural tradition of great power and imagination was developed by a society that was completely isolated from external cultural influences of any kind for over a millennium. Rapa Nui, the indigenous name of Easter Island, bears witness to a unique cultural phenomenon. A society of Polynesian origin that settled there c. A.D. 300 established a powerful, imaginative, and original tradition of monumental sculpture and architecture, free from any external influence. From the 10th to the 16th century this society built shrines and erected enormous stone figures known as moai, which created an unrivalled landscape that continues to fascinate people throughout the world. Rapa Nui National Park is a testimony to the undeniably unique character of a culture that suffered a debacle as a result of an ecological crisis followed by the irruption from the outside world. The substantial remains of this culture blend with their natural surroundings to create an unparalleled cultural landscape.

These criteria are entirely cultural and include not only the assessments that justify the creation of the World Heritage Site but also chronological data and interpretative assertions that are still objects of scientific debate and will be discussed throughout the book. Nevertheless, whatever the explanation that better fits with the available evidence, the uniqueness and significance of Easter Island's cultural heritage is beyond all doubt. The lack of criteria aimed at conserving nature is probably due to the already mentioned low biodiversity and the full degradation that the island's landscape and its natural systems have suffered since its human settlement. In its present status, the Rapa Nui National Park has an extension of $71\,\mathrm{km}^2$, which represents 43% of the total island surface (Fig. 1.18), and is ruled by the National Forestry Commission (CONAF) of the Chilean Ministry of Agriculture. According to the current management plan, the main objectives of the park are (i) to protect and conserve the Rapanui archaeological and historical heritage, including cultural manifestations such as language, beliefs, and traditions; (ii) to protect and recover the native flora and the flora introduced by Polynesian colonizers; (iii) to conserve coastal marine ecosystems; (iv) to control factors of environmental deterioration (erosion, fire, plagues, contamination); (v) to secure a solid and permanent economic activity based on cultural tourism; and (vi) to support scientific research, provided it fits with the conservation of natural and cultural heritage, as well as the needs and feelings of the local community (CONAF, 1997). Several initiatives developed under the auspices of the Rapa Nui National Park, with the collaboration and funding of international organisms, have been advanced to conserve and/or restore the archaeological and natural heritage of Easter Island.

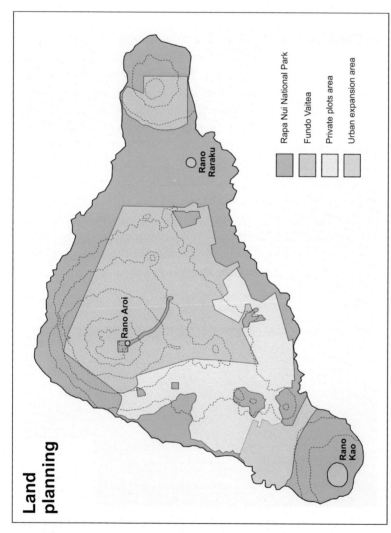

Land planning

- Rano Aroi
- Rano Raraku
- Rano Kao

Rapa Nui National Park
Fundo Vaitea
Private plots area
Urban expansion area

Fig. 1.18 Map of land planning for Easter Island. Fundo Vaitea is part of the former sheep ranch (Section 1.4.3), which is presently closed to the general public. The private plots area is where most cultivation and cattle raising activity develop (see Fig. 1.9). *(Redrawn from Vargas, P., Cristino, C., Izaurieta, R., 2006. 1.000 años en Rapa Nui. Cronología del asentamiento. Editorial Universitaria, Santiago de Chile and CIREN, 2013. Delimitación Parque Nacional. Centro de Información de Recursos Naturales, Ministerio de Agricultura, Santiago de Chile. Available at https://www.ciren.cl/; Accessed 6 November 2019.)*

1.6.1 Conservation of archaeological heritage

When the Rapa Nui National Park was created, most ahu were totally or partially destroyed and most moai were toppled (Fig. 1.19). This has been attributed to several causes by different authors, including earthquakes and social conflicts, eventually warfare, which may have contributed to intentional moai toppling and fragmentation among rival clans (Sections 2.4.2 and 2.4.3). The reconstruction of these monuments began by the mid-20th century and, to date, about 50 moai have been reerected on their ahu (Hunt and Lipo, 2011). The North American archaeologist William Mulloy, who dedicated more than two decades to the research and restoration of Easter Island's archaeological heritage, deserves special mention. Significant preservation actions include the reconstruction of several ahu (Akivi, Vai Teka, Ko Te Riku, Tahai, Vai Uri, Uri A Urenga, Hanga Kio'e, O Kava) (Fig. 1.1) and the ceremonial village of Orongo between 1960 and 1978 (Mulloy, 1970b, 1973, 1975; Mulloy and Figueroa, 1961, 1978; Figueroa, 1979). The World Monuments Fund (WMF) was instrumental for the development of these tasks (https://www.wmf.org/project/easter-island-rapa-nui%E2%80%94moai-conservation-project). The action of tsunamis, although sporadic, may be devastating for the moai (Margalef et al., 2018). For example, in 1960, the Valdivia earthquake—centered on

Fig. 1.19 A destroyed ahu and a toppled moai near Ahu Akahanga (Fig. 1.1). *(Reproduced under CCA 2.0 Generic license. Image source: https://commons.wikimedia.org/wiki/File:Easter_Island_moai_near_Ahu_Akahanga_(6680552429).jpg#filelinks.)*

the continental Chilean coasts at about 3700 km of Easter Island—generated a tsunami that reached the island in the form of tidal waves up to 8 m high. This destroyed many ahu from the SE coasts, including Ahu Tongariki (Fig. 1.13), whose remains were spread over more than four hectares and some moai were broken and pushed more than 150 m inland (Cristino and Vargas, 1999; Cortez et al., 2009). Ahu Tongariki was rebuilt to its present aspect between 1992 and 1994 at the initiative of Chilean archaeologists Caludio Cristino and Patricia Vargas, with the help of international partners, notably the private Japanese company Tadano Corporation (Cristino and Vargas, 1999, 2002).

Another threat to moai integrity is the decay that may cause the long-term exposure to atmospheric agents. The tuff of which the moai are made is a soft and porous rock (Gioncada et al., 2010) and is susceptible to mechanical damage during their carving, transportation, and emplacement. In the long run, the moai are also subjected to deterioration, particularly under the frequent rain showers causing surface abrasion and penetration within its porous matrix, which favors mineral dissolution and chemical weathering. Sudden changes of temperature can also contribute to rock deterioration by recurrent expansion and contraction facilitating rock breakage. The growth of algae and lichens on the moai surface (Follmann, 1961) may also enhance rock weathering by retaining water and dust, as well as by lichenic hyphae penetration (5–10 cm) into the rock fissures and porosity, thus causing mechanical and chemical damage (Fig. 1.20). Activities such as warfare, vandalism, and the introduction of exotic animal species may have also accelerated disintegration. Toppling and cracking increases the attack surface, and hence the exposure to physicochemical and biological weathering. The extensive exploitation of Easter Island as a sheep ranch also contributed to the deterioration of moai, especially those broken and lying on the ground, which were used as rubbing posts. To a lesser extent, the same occurs today with cattle and horses. Materials from several ahu were also used to build stone fences and other structures (Cristino and Vargas, 1999). All these decay processes and their physical, chemical, and biological mechanisms were summarized by Charola (1997) and Sawada et al. (2001). Petroglyphs are also submitted to deterioration, notably by animal damage (quadruped trampling, bird droppings, and tunneling insects), plant growing (algae, moss, lichens, grass), erosion, graffiti, and abrasion (Charola et al., 1998).

According to Charola (1997), in addition to the reconstruction of as many ahu as possible, the remedy for the moai should address two main points, namely, strengthening the stone and preventing water penetration.

Fig. 1.20 Example of a *moai* from Ahu Tongariki affected by surface weathering due to the exposure to atmospheric agents. Lichens growing on the rock are also visible as clear spots. *(Photo: V. Rull.)*

As a pilot essay, a water-repellent treatment was applied to the moai at Hanga Kio'e, near Hanga Roa (Fig. 1.1), after a careful removal of biological and chemical deposits from the surface (Roth, 1990). This was carried out by the National Center of Conservation Center of Santiago, with support from UNESCO and the private company Wacker Chemie. The treatment was proven to be effective and it was considered that it should be reapplied every 5–15 years. Charola (1997) pointed out that the only way of stopping deterioration is to install the moai permanently in protected sites such as museums but, in the words of the same author, "…moai in a museum no longer lives: it is not on the ahu, it cannot hear the sea or feel the sun, it can no longer watch the spectacle presented by the descendants of his creators. Is it solution worth it?" It could be added that the moai only make sense if they are preserved in situ, as the "living" legacy of the prehistoric Rapanui culture to its descendants and the witnesses of its historical developments and the future that awaits them. In other words, the outdoor museum of Easter Island is the right place for the moai and their future should be on the hands of the Rapanui descendants. To preserve the archaeological heritage of Easter Island, in general, Charola et al. (1998) emphasized the need for understanding its value, especially by the local population, and establishing an effective system of surveillance.

1.6.2 Conservation of native and Polynesian-introduced flora

In spite of the evident degradation of Easter Island's flora, vegetation, and landscape, it is clear that the park administration has genuine interest in nature conservation, with emphasis on native plant species and others of cultural interest, mainly those introduced by Polynesian settlers. The identity of these species is usually found in the reports of the first European visitors or is deduced from the present distribution and the dispersal mechanisms of the involved species (Zizka, 1991), but is also discovered after the study of charcoal remains from archaeological sites (Orliac, 2000; Orliac and Orliac, 1998) and palynological analysis of lake and swamp sediments (Flenley et al., 1991; Cañellas-Boltà et al., 2014; Horrocks et al., 2015) (Section 1.4.3).

Several conservation and restoration efforts have been done by CONAF in collaboration with ONF International, a subsidiary of the French National Forestry Office (ONF). Among them, the program Umanga mo te Natura, initiated in 2005 with the participation of the Rapanui community, was aimed at preserving and restoring the native flora on the basis of its remaining relics, stopping the advance of recently introduced invader species and controlling soil erosion by recuperating the plant cover of eroded lands. In this case, invasive species, grazing and agriculture expansion were also considered relevant threats to the native flora (Dubois et al., 2013). These actions were carried out in several sites of the island, including the inner slopes of Rano Kao and Rano Raraku, and several species considered to be native were selected to be planted, including the angiosperms *Sapindus saponaria, Triumfetta semitriloba, Sophora toromiro* and the ferns *Cyclosorus interruptus* (= *Thelypteris interrupta*), *Vittaria ensiformis* (= *Haplopteris ensiformis*), *Asplenium polyodon,* and *Microsorum parksii.* Other species planned to be restored were *Tetragonia tetragonoides, Apium prostratum, Chenopodium glaucum, Lycium sandwicense* (= *Lycium carolinianum* var. *sanwicense*), and *Paspalum forsterianum* in the Ovahe beach (Fig. 1.1). Among the invader species to be removed, *Melinis minutiflora* was especially noteworthy. The recovery of eroded lands was focused on the Poike peninsula as a pilot area, and the plants used for repopulation were several alien woody species already growing on the island (*Dodonaea viscosa, Casuarina equisetifoila, Albizia lebbeck,* and *Thespesia populnea*). The results of these actions have not yet been published and it is difficult to know whether or not they have been successful.

In the specific case of the toromiro, Maunder et al. (2000) mentioned several unsuccessful attempts to reintroduce this species on Easter Island from the specimens growing on botanical gardens abroad. Dubois et al. (2013) attributed this failure to the lack of the original species' niche, whose characteristics are unknown and have most probably been removed by human action. Espejo and Rodriguez (2013) and Püschel et al. (2014) addressed the efforts made to reintroduce *Sophora toromiro* in Easter Island from specimens growing in botanical gardens located in continental Chile (Viña del Mar and Titze) and Sweden (Göteborg). The main activities were propagation by grafting using rootstocks of similar species of the same genus (*Sophora cassioides*) and seed germination from herbarium specimens. This material should be the basis for an orchard/arboretum that should provide the plants for active reintroduction in the medium term (3–5 years) (Espejo and Rodriguez, 2013).

It should be highlighted that restoration of preanthropogenic palm forests is unfeasible as the palm species that dominated these forests was likely endemic to the island and is already extinct. The use of a similar palm species would be totally unnatural. In the first place, it would be necessary to find a species with the same ecological requirements as the extinct palm, which are unknown. In addition, it is doubtful that the ecological relationships of the eventual substitute palm with other preanthropic forest species were maintained and the interactions with other members of the ecosystem were still possible. Building an ecosystem is not only a matter of assembling a set of populations of different species. Ecosystems develop and establish by ecological succession, which is a long and unpredictable process without a definite target (West et al., 1981). Ecological engineering is still unable to initiate and control a process like this to achieve a premeditated output. It should also be taken into account that soils have been highly degraded after the initial deforestation and the subsequent extensive grazing, and <40% of the island surface seems to be able to support forested vegetation (Section 1.3). In addition, the socioeconomic impact of an eventual large-scale reforestation program of the island should also be considered, as the necessary investment would be very high and it is not sure that such a venture may be in line with the present-day islander's needs and feelings.

CHAPTER 2

The prehistory: From human settlement to European contact

Contents

2.1	From exploration to research	42
2.2	Prehistoric chronology	44
2.3	The first settlers	49
	2.3.1 Some terminological considerations	49
	2.3.2 When and from where?	50
	2.3.3 From Polynesia to America and back	55
	2.3.4 Summary	59
2.4	The ancient Rapanui society	60
	2.4.1 Sociopolitical organization	60
	2.4.2 The moai cult	63
	2.4.3 The Birdman cult	67
	2.4.4 Deforestation	72
	2.4.5 Collapse or resilience?	74
	2.4.6 Demography	80
	2.4.7 The genocide	83
	2.4.8 Summary	85
2.5	A narrative of human determinism	87

This chapter is a summary of the main cultural developments of the Easter Island's prehistory, based on archaeological, anthropological, ethnographical, ethnobotanical, and historical evidence. The main aim is to provide a sociocultural framework for paleoecological trends and events to be disclosed in further chapters. A number of relevant prehistoric issues and their particular chronological boundaries are still under discussion by specialists. Examples are the timing of initial settlement, the subdivision into cultural phases or periods, the chronology and the causes of the shift from the moai cult to the Birdman cult, the timing, progress and causes of total forest clearing, or the demographic trends of the ancient Rapanui society, among others. This brief account by a nonspecialist is not an attempt to discuss these

Paleoecological Research on Easter Island
https://doi.org/10.1016/B978-0-12-822727-5.00002-7

topics but a trial to review the existing literature in the search for eventual agreements or coincidences. In those matters where a consensus is still to be reached, the existing proposals are presented and the most relevant literature is provided. In this and the rest of the chapters, the following chronological notation will be used:

- BCE—Before Common Era (formerly BC)
- CE—Common Era (formerly AD)
- ^{14}C (k)yr BP—Radiocarbon (kilo)years before present
- Cal (k)yr BP—Calibrated (kilo)years before present

2.1 From exploration to research

Archaeological and ethnographic research began about a century and a half after European arrival. The first European explorers—i.e., the Dutch Jacob Roggeveen (1722), the Spanish Felipe González de Haedo (1770), the English James Cook (1774), and the French Jean-François Galaup, Count of La Pérouse (1786)—and some members of their crews recorded a variety of random observations about the natural history and the culture of the island in their logbooks (Fischer, 2005; Richards, 2008) (Fig. 2.1). After this exploratory phase, the island experienced a turbulent epoch in which adventurers, whalers, traders, and slave raiders significantly altered the life of the Rapanui and drastically decimated its population (Section 2.4.7) (Richards, 2017). This has been called the "Dark Era" by Boersema (2015a). The first more or less systematic archaeological observations were published a century and a half after European contact by the British surgeon John L. Palmer, who described and mapped a few ahu (Palmer, 1870). This occurred at the end of slave raids and the beginning of the extensive exploitation of the island as a sheep ranch (Section 1.4.3) when the Rapanui population attained its minimum (a few more than a hundred individuals) and was about to disappear due to slavery and alien epidemic diseases. This event is also known as the genocide of the Rapanui society (Section 2.4.7). A couple of decades later, when the island had already been annexed by Chile, the North American Navy paymaster William J. Thompson carried out the first comprehensive survey of the island, describing and mapping more than 100 ahu and 500 moai (Thompson, 1891). The British historian Katherine Routledge spent 16 months studying the island and published a systematic inventory of moai and ahu, along with a compilation of traditional stories, language features and customs, obtained by oral transmission from the Rapanui people of that time (Routledge, 1919). A couple of decades later,

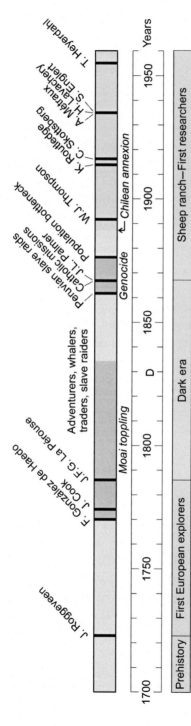

Fig. 2.1 Main historical trends and events on Easter Island since the European contact to the arrival of the Norwegian Heyerdahl's expedition. In the upper bar, the time during which Easter Island was politically independent is indicated in *yellow* and the time after Chilean annexation, in *blue*. The *pink* intervals correspond to the phase of moai toppling and the postcontact genocide (i.e., population demise and acculturation) of the Rapanui society (Section 2.4.7). In the same bar, vertical *blue lines* indicate the arrival of the explorers and researchers mentioned in the text and some milestone historical events for Easter Islanders. (*Raw data from Fischer, S.R., 2005. Island at the End of the World. The Turbulent History of Easter Island. Reaktion Books, London and Boersema, J.J., 2015a. The Survival of Easter Island. Dwindling Resources and Cultural Resilience. Cambridge University Press, Cambridge.*)

the Swiss anthropologist Alfred Métraux and the Belgian archaeologist Henri
Lavachery expanded the work of Routledge, with emphasis on ethnography
and petroglyphs, respectively (Lavachery, 1939; Métraux, 1940). The catholic
priest Father Sebastian Englert arrived in 1935 and remained on the island for
more than 30 years, until his death in 1969. He is known for his studies on the
Rapanui language and ethnography, as well as for his archaeological contributions in the form of inventories of moai and pukao (Englert, 1948, 1970).

The 1955–56 Norwegian expedition led by the explorer Thor Heyerdahl
represented a milestone in the study of Easter Island and laid the foundations of modern archaeological research (Ayres and Stevenson, 2000; Smith,
2000; Hunt and Lipo, 2018). However, this influence was not due to the
ideas of Heyerdahl himself (Section 2.3.2) but to the fact that the expedition included several archaeologists and natural scientists that continued
their own research on Easter Island and greatly influenced further generations. In terms of archaeological research, the most influencing scientist
of the Heyerdahl's team was William Mulloy, who has already been mentioned in the section about the restoration of the archaeological heritage
(Section 1.6.1). The other important scientist of this expedition was the
Swedish botanist Carl Skottsberg, who completed the study of the island's
flora initiated by himself in 1917 (Skottsberg, 1920–1956). During the last
decades of the 20th century, detailed studies of specific areas and/or particular archaeological features intensified (Hunt and Lipo, 2018) but a detailed
account is beyond the scope of this book. Only the main traits of this research are presented, subdivided into three sections: chronology, settlement,
and cultural developments of Easter Island's prehistory.

2.2 Prehistoric chronology

The first chronological subdivision of Easter Island's prehistory was developed during the Heyerdahl's expedition and considered three main
periods, namely the Early Period (400–1100 CE), the Middle Period (1100–
1680 CE), and the Late Period (1680–1722 CE) (Fig. 2.2), which were
based on sequential changes in the ceremonial architecture (Smith, 1961).
The Early Period or Ahu Moroki was defined as the time of the establishment of the Polynesian settlers on the island and the building of the first,
rather rudimentary, ahu and moai. During the Middle Period, also called
the Expansion Period or the Ahu Moai Period, the human population increased, the ahu were significantly improved, and the moai grew in number
and size. During the Late Period, also known as the Decadence Period or

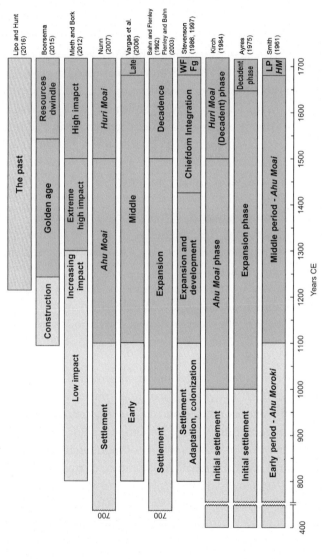

Fig. 2.2 Selection of the most representative chronological frameworks for the prehistory of Easter Island, sorted by publication dates (bottom to top). The three phases/periods with similar, but not identical, cultural patterns have been highlighted with different colors, except the latest published "chronology," which is in *gray*. Note the strong disagreement in the different dates for island settlement and in the boundaries among the different phases/periods defined. Fg. Moai fragmentation; HM, Huri Moai; LP, Late Period; WF, Warfare.

the Huri Moai Period, the moai industry ceased, most of the existing moai were toppled and the ahu were destroyed. In spite of the low amount of radiocarbon dates supporting the boundary ages, this temporal framework and the evidence that supported it was the first chronological attempt and, albeit criticized (Golson, 1965), constituted the reference for Easter Island research during decades (Vargas et al., 2006; Mulrooney et al., 2009). Some modifications to these chronologies were introduced by Ayres (1975) and Kirch (1984), who placed the beginning of the Ahu Moai phase at 1000 CE. In addition, Kirch situated the start of the Huri Moai phase at 1500 CE, roughly a century and a half before its original position (Fig. 2.2). Stevenson (1986) returned to the original chronology of Smith but with some variations. He modified the date of original settlement to 800 CE and added a phase of sociopolitical reorganizations, called Chiefdom Integration, before the Huri Moai phase that this author called Warfare and Moai Fragmentation phase. Bahn and Flenley (1992) and Flenley and Bahn (2003) adopted the framework of Kirch (1984), with an earlier settlement date (700 CE), and recovered the names previously given by Ayres (1975). These authors also proposed alternative names for the three phases, such as Altars, Statues, and Burials or Architecture, Sculpture, and Rock Art. They also suggested that the moai would have been toppled after European contact, rather than in prehistoric times, as proposed by the first chronologies. Vargas et al. (2006) came back to the first subdivision into Early, Middle, and Late periods, as defined by Smith (1961), albeit they shortened the Early Period by adopting the Polynesian arrival at 800 CE. Nunn (2007) used the same chronology as Kirch but modifying the settlement date (Fig. 2.2).

Other chronologies combined cultural and environmental aspects. For example, Mieth and Bork (2010), defined three main phases according to the Rapanui demography and its impact on island ecosystems. According to these authors, after settlement (800 CE), a low-impact phase was characterized by the growing population, the beginning of monumental architecture (i.e., the moai and the ahu) and small-scale cutting of trees (Fig. 2.2). Intense forest clearance started by 1250–1350 CE and was the onset of a phase of extreme environmental impact, with high population densities and the acme of monumental architecture. This phase would have extended until 1450–1550 CE when the island was almost deforested and the Rapanui, still very numerous, shifted to more extensive agricultural practices and the monumental architecture continued active until de European contact. A recent attempt to synthesize the cultural and environmental chronology of prehistoric Easter Island was made by Boersema (2015a), who considered that Easter Island was discovered and settled by humans by 1100 CE (Fig. 2.2). From 1100 to 1250 CE (Phase I or Construction Phase) the colonizers would have started to establish on the island by building the necessary infrastructures to live and the first moai for cult purposes. During this phase, low-intensity tree felling and burning would have occurred. Phase II (1250–1550 CE) is considered in this framework the "Golden Age" of the ancient Rapanui society, characterized by maximum population estimates and the peak of the moai industry and cult. At the same time, large-scale island deforestation would have been in progress. The intensive and extensive exploitation of natural systems would have led to Phase III (1550–1722 CE), characterized by the total disappearance of forests and a general resource dwindling causing a population decline. At the same time, the moai cult gradually disappeared and was replaced by a new religious practice known as the Birdman cult. Some of the discussed chronologies are explicitly linked to the deforestation process, which is an integral part of the temporal subdivision, commonly associated with the cultural Rapanui expansion or Ahu Moai phase. However, for a sound deforestation chronology, all the paleoecological evidence available to date is needed (Chapter 7), rather than only the pioneering palynological records (Chapter 4), which are the most used in the above chronological frameworks.

A quick look into Fig. 2.2 shows that most chronologies define three phases or periods that could be summarized as follows: (i) the first phase of settlement with low population density and limited environmental impact, which represented the initiation of the moai industry; (ii) the second phase of cultural expansion with higher population density, intense architectural expansion and high levels of environmental impact; and (iii) the third phase of decadence, characterized, in some cases, by resource depletion, cultural degradation, and monument destruction. However, in spite of this more or less general coincidence in prehistorical trends, there is a strong disagreement in the age of these phases/periods and also in the settlement dates. Therefore, a robust chronology of human developments on Easter Island is still to be defined. In the words of Lipo and Hunt (2016):

> "Despite over a hundred years of research and at least 500 radiocarbon samples that span about 60 years of collection our knowledge of the prehistory of Rapa Nui is just beginning. Our knowledge is limited because archaeologists have yet to generate a systematic set of measurements allowing us to specify the timing of human events with any real precision. In general, our most precise knowledge consists of the timing of the arrival of humans to this tiny, remote island in the 13th century and the subsequent arrival of European explorers that begins AD 1722." ... "Beyond

the knowledge of these events, the archaeological record can only largely be dated to 'the past' with little more precision than that" (Fig. 2.2).

It could be added that the situation is even worst as the date of arrival of Polynesians to Easter Island is still disputed, as discussed in Section 2.2. Therefore, it seems that, according to this view, the only chronological certainty is that the first Europeans arrived in 1722 CE, which is known by historical documents. For Lipo and Hunt (2016), this situation is due to the general "…lack of rigor in requiring a firm relation between the dated and target events…," which has derived in the accumulation of radiocarbon dates with poor systematic value, in terms of archaeological chronology. For these authors, the solution is to "… systematically identify samples [for dating purposes] that have a clear and robust relation to events of interest." It should be noted, however, that this is not an exclusive problem of Easter Island's archaeology, as it can be extended to scientific research, in general. The recent spectacular increase of computational power has made possible the creation and management of huge databases of public access. However, this impressive amount of information is useless if there is not a specific research purpose or, in other words, if potential users do not have specific questions to ask the data. Databases and data-mining software applications cannot design and manage, by themselves, intelligent research addressed to specific targets of interest.

In the absence of a generally accepted prehistoric chronology, the following sections are not sorted chronologically but grouped according to cultural events and concepts such as settlement, cult practices or demography, among others. The review is based on the available archaeological, anthropological, ethnographic, ethnobotanical, and historical literature with no reinterpretations. A caution call is opportune in the case of ethnographic evidence. As already mentioned above (Section 2.1), and will be explained in more detail in Section 2.4.5, the Rapanui society experienced a population bottleneck at the end of the 19th century, which left the transmittal of the Rapanui tradition in the hands of barely a hundred individuals belonging to a few families (Section 2.4.7). This would have constrained and/or biased the oral transmission of Rapanui history, legends and traditions (Flenley and Bahn, 2003). A detailed and updated review of the Rapanui life and beliefs based on oral tradition is available at Edwards and Edwards (2013). The following account of human developments is only for orientation purposes, with the aim of setting a scenario of cultural events where to fit the available paleoecological findings, to which the rest of the book is devoted.

2.3 The first settlers

2.3.1 Some terminological considerations

Before starting, some terms related to colonization may require some clarification, to facilitate further discussions. Lipo and Hunt (2016) differentiated between the intuitive concept of "colonization," as an imagined set of events related with living in a new land, from "archaeological colonization," which is an account of the earliest observations that can only be explained as a result of human activity and involves the seeking of chronological information for the deposition of the earliest artifactual material. The concept of archaeological colonization is adopted in this book and other terms and concepts are added. According to the Cambridge Dictionary (https://dictionary.cambridge.org/) the term "discovery" means "finding something or someone for the first time" and the term "settlement" means "a town or village that people build to live in after arriving from somewhere else." Note that discovery does not imply the establishment of human populations in the discovered site, whereas settlement involves effective human occupancy of the site. Three types of the settlement will be distinguished, namely ephemeral settlement (if people abandon the settled site), intermittent settlement (when people, either of the same or of different origins, recurrently establish and abandon the site) and permanent settlement or colonization (when settlers adopt the settled site as their permanent living place and develop their own culture). If a settlement can be documented archaeologically, that is, by artifactual material as evidence of human activity, then it coincides with archaeological colonization (*sensu* Lipo and Hunt, 2016). This terminological scheme is introduced only for the reader to be aware of the meaning in which the terms are used throughout the book and should not be considered either a proposal for a new terminology of general validity or a counterproposal to other conceptual frameworks (e.g., Weisler and Green, 2011).

Discovery is an event that can easily be unnoticed because of the lack, scarcity, or further removal of evidence of human presence. Sometimes, however, discovery events can leave some signal of human activity. For example, in the Azores Islands (Atlantic Ocean), evidence for human discovery without permanent settlement was detected from the occurrence of the first fires along with the presence of spores from coprophilous fungi (notably *Sporormiella*)—which are indicators of the presence of domestic animals (Van Geel et al., 2003)—around 1100 CE, three centuries and a half before Portuguese colonization (1449 CE). These domestic animals were left on the wild and inhabited islands—which, by those times, were devoid of terrestrial mammals—by sailors usually passing by, but not living in, the archipelago to guarantee a continued food supply during their travels. More than a century and a half later (1270 CE), the first signs of human settlement were recorded by a slight fire increase and the finding of pollen from rye (*Secale cereale*) and other cereals, which represented the onset of small-scale agricultural practices on the islands. However, these agricultural records were discontinuous in time, suggesting itinerant and/or intermittent settlement. Large-scale, permanent land settlement (i.e., colonization) began by 1400 CE and was manifested in full deforestation, a significant increase of fires and the intensification of agricultural and forestry practices. The details of the whole process can be found in Rull et al. (2017) and De Boer et al. (2019). These findings were based on paleoecological analyses of lake sediments, archaeological remains were absent during the discovery (early 12th century) and early settlement (mid-to-late 13th century) phases. Therefore, archaeological colonization would not have occurred until the mid-15th century but humans, of unknown origin, had already discovered and settled the Azores Islands some centuries before.

2.3.2 When and from where?

The initial settlement of Easter Island by humans is one of the more controversial issues, both for the time of arrival of the first settlers and for their geographical origin, either Polynesia, to the West, or South America, to the East (Fig. 1 of the Introduction). The topic has recently been reviewed by Rull (2019), which constitutes the basis for this section.

Regarding the place of origin, the first ethnologists working on the island suggested that the legendary island of Hiva—from where, according to the Rapanui oral tradition, the first Polynesians sailed to Easter Island—could have been located in the Marquesas or the Gambier Islands (Fig. 2.4), and the best candidate seemed to be the Gambier's Mangareva Island (Routledge, 1919; Métraux, 1940). These authors were unable to deduce any reliable colonization date from the Rapanui legends. Several years later, Thor Heyerdahl proposed a radically different hypothesis, according to which Easter Island would have been settled from South America by Amerindian cultures. This hypothesis was based on the observation that the predominant winds and marine currents favored E-W navigation without the need for sophisticated technologies. To demonstrate this hypothesis, Heyerdahl organized the famous Kon-Tiki expedition that, in 1947, navigated from the South American coasts of Perú to the Tuamotu Islands (Fig. 2.4) in 101 days, using a rudimentary raft similar to those used by aboriginal Americans (Heyerdahl, 1952). For Heyerdahl, this was the

Fig. 2.3 The raft used by Heyerdahl and his colleagues in the Kon-Tiki expedition from Perú to the Tuamotu Islands. *(Reproduced under license CCA 2.0 G from https://commons. wikimedia.org/wiki/File:Expedition_Kon-Tiki_1947._Across_the_Pacific._(8765728430).jpg.)*

demonstration that Amerindians would have been able to arrive in Easter Island with the simplest possible navigation technology (Fig. 2.3).

A few years later, Heyerdahl organized the well-known 1955–56 field-work campaign to Easter Island in the search for evidence supporting his idea. He found, for example, that some plants, notably the sweet potato (*Ipomoea batatas*), were considered to be introduced from South America and their names (kumara, in the case of sweet potato) were also of American origin. He also noted similarities between many tools and constructions, including the moai and some petroglyph figures, with potential Incaic and pre-Incaic counterparts from South America (Heyerdahl and Ferdon, 1961). The underlying hypothesis of Heyerdahl was that Native Americans were the discoverers and firsts settlers of Easter Island by approximately 400 CE and the Polynesians arrived later and eradicated the Amerindian culture. Heyerdahl's chronology was based on a single radiocarbon date of 1570 ± 100 ^{14}C yr BP (386 CE) that was called into question later and considered too old in comparison to many other dates from archaeological materials of the same provenance (Martinsson-Wallin and Crockford, 2002). Most evidence provided by Heyerdahl to defend his theory was re-analyzed by Flenley and Bahn (2003) to support the contrary hypothesis. These authors highlighted that the winds and currents that favor E-W navigation are not constant but vary seasonally and that the ENSO phenomenon may periodically revert winds and currents thus favoring W-E navigation. Furthermore, Polynesians were experienced sailors who have colonized most Pacific islands and archipelagos for millennia in a W-E direction, using highly hydrodynamic and efficient boats moved by wind and oars. Regarding plants of South American origin and their names, these would have been introduced after European contact (1722 CE), occurred 230 years after the European discovery of America (1492 CE). According to Flenley and Bahn (2003) the Rapanui art—including the moai and the tools used for their carving, along with the images of the petroglyphs—was similar to that of other Polynesian islands and archipelagos. Also, the skull biometry and the dental patterns of skeletons from ancient Rapanui found in Easter Island excavations were clearly of Polynesian origin (Pietrusewsky and Ikehara-Quebral, 2001; Stefan, 2001).

With all this evidence, Flenley and Bahn (2003) dismissed Heyerdahl's hypothesis and concluded that Easter Island was discovered, settled, and colonized only once and from Eastern Polynesia. According to these authors, forests started to decline by 800 CE (Chapter 4), which could be close to the date of Polynesian settlement, under the hypothesis of anthropogenic deforestation. Using archaeological evidence, other authors suggested that the initial settlement could have occurred between 800 and 1000 CE, and the ceremonial sites with worked stones were erected later, between 1100 and 1200 CE (Steadman et al., 1994; Martinsson-Wallin and Crockford, 2002; Vargas et al., 2006). Regarding the place of origin, Martinsson-Wallin and Crockford (2002) expanded the geographical range to the Mangareva-Pitcairn-Henderson area or the Tuamotu Islands (see also Kirch and Green, 1987; West et al., 2017). This chronology is consistent with the more accepted chronology of colonization of the Pacific islands and archipelagos from eastern Asia. In this framework, the colonization of Easter Island was one of the last steps of a W-E spreading process initiated by 5000 yr BP (3000 BCE), when Taiwanese sailors colonized the Philippines (Kirch, 2010) (Fig. 2.4). The next step was the colonization of the Bismarck Islands, north of New Guinea, by 1500 BCE. This was a fundamental stage because it implied the disconnection from the original Taiwanese culture and the development of a new culture, known as Lapita, which was the seed of all Polynesian cultures. The Lapita culture experienced a rapid expansion reaching Tonga and Samoa (Western Polynesia) by 950 BCE, followed by a long pause of approximately 2000 years, before the expansion toward Eastern Polynesia. The archipelagos of Society, Tuamotu, and Marquesas, as

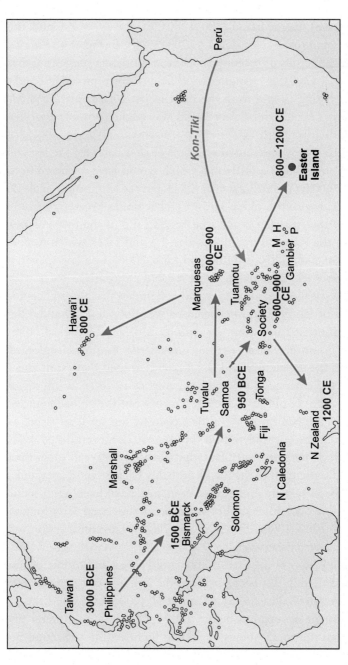

Fig. 2.4 Peopling of the Pacific from East Asian archipelagos. H, Henderson Island; M, Mangareva Island; P, Pitcairn Island. The approximate trajectory of Heyerdahl's Kon-Tiki expedition is also indicated *(green arrow)*. *(Raw data from Kirch, P.V., 2010. Peopling of the Pacific: a holistic anthropological perspective. Annu. Rev. Anthropol. 39, 131–148.)*

well as the Mangareva Island, were not colonized until 600–900 CE. The last expansion wave started from these archipelagos in three different directions: Hawaii (800 CE), Easter Island (800–1200 CE), and New Zealand, the last Polynesian Island to be colonized, by 1200–1300 CE (Fig. 2.4). This is consistent with the results of recent paleoclimate modeling suggesting that between 800 and 1300 CE, changing wind field patterns provided conditions in which voyaging to and from the most isolated East Polynesian Islands, such as New Zealand and Easter Island was readily possible by off-wind sailing (Goodwin et al., 2014).

Some authors have presented evidence of later dates for the last colonization wave, including Easter Island, which would have occurred between 1200 and 1300 CE (Wilmshurst et al., 2011; but see Mulrooney et al., 2011). This coincided with the oldest known archaeological materials from Ahu Nau Nau, in the Anakena beach (the landing site of the first Polynesians, according to the Rapanui tradition) (Figs. 1.1 and 1.11), which were dated between 1200 and 1400 CE (Hunt and Lipo, 2006). In a later paper, these authors made a selection of the most reliable radiocarbon dates with errors of < 10% of the measured ages and concluded that the archaeological colonization of Easter Island occurred between about 1220 and 1260 CE (Lipo and Hunt, 2016). Given the definition of archaeological colonization by these authors, these ages are necessarily minimum colonization ages (Section 2.3.1). Flenley and Bahn (2011) argued that Hunt and Lipo (2006) used the absence of evidence as evidence of absence and suggested that colonization would have occurred earlier but the archaeological evidence could have been flooded or destroyed by the sea level rising experienced during the last millennium (Nunn, 2007). Stevenson et al. (2000) considered that a date of 1100 CE is approximately 300 years older than the hypothesized settlement of the island, probably because most areas of Easter Island have been heavily eroded and contain few archaeological contexts sealed by sediments. Other studies on Anakena suggested early human activity during 1000–1300 CE, before Ahu Nau Nau construction, which would have occurred between 1300 and 1450 CE (Martinsson-Wallin and Wallin, 2000; Wallin et al., 2010). This could be an indication of human settlement (as used in this book) prior to the building of ceremonial structures (Martinsson-Wallin, 1998). Mann et al. (2003) suggested the possibility of transient occupation before 1200 CE by hunter-gatherer people who exploited the rocky shores, including flightless birds (Section 5.2.2). In the terminology of this book, this could correspond to ephemeral or intermittent settlement (Section 2.3.1).

2.3.3 From Polynesia to America and back

The story of Easter Island colonization seems not to be as simple as a dual and exclusive debate between Polynesia and South America, as the places of origin of the first settlers. In chronological terms, the situation is similar, as a single colonization event from either one or the other side has recently been called into question. The disturbing element is the sweet potato or kumara (Fig. 1.10). This species originated in tropical America and was domesticated more than 6000 years ago, owing to the nutritional properties of its tuberous roots (Piperno and Holst 1998; Bovell-Benjamin, 2007). The oldest Polynesian record of sweet potato consisted of carbonized soft tissues found on the Cook Islands (Fig. 1 of the Introduction), dated to approximately 1000 CE (Hather and Kirch, 1991). The occurrence of the sweet potato remains throughout Polynesia in archaeological sites of similar ages has suggested that cultivation and consumption of this plant was common and widespread long before European contact. On Easter Island, there is sound evidence of sweet potato cultivation and consumption long before European contact, even before Europe-America contact. Indeed, the microfossil analysis of archaeological sites demonstrated the presence of sweet potato crops by 1300 CE (Horrocks and Wozniak, 2008). Also, carbonized fragments of sweet potato were found on archaeological sites dating back to 1–3 centuries before European contact (Orliac and Orliac, 1998; Orliac, 2000). In addition, consistent presence of starch grains of *Ipomoea batatas* has been found in the dental calculus of Rapanui skeletons dated to 1330 CE, about four centuries before the arrival of Europeans to the island and a century and a half before the European discovery of America (Tromp and Dudgeon, 2015). It has been considered that the sweet potato was the main food source during the flourishment of the ancient Rapanui society between approximately 1200 and 1500 CE (Fischer, 2005; McCall, 2009). Four hypotheses have been proposed to explain the presence of sweet potato on Easter Island before European arrival in America; these are the long-distance dispersal hypothesis, the back-and-forth hypothesis, the Heyerdahl hypothesis, and the newcomers hypothesis (Fig. 2.5).

According to the long-distance dispersal hypothesis, sweet potato seeds could have reached Easter Island by nonhuman means (Flenley and Bahn, 2003), including birds, wind, or rafting (attached to sea-drifting materials). Dispersal modeling suggested that the probability of long-distance seed dispersal from America and further successful establishment in Polynesia was very low (Montenegro et al., 2008). However, recent molecular phylogenetic studies seem to suggest the long-distance dispersal across the Pacific

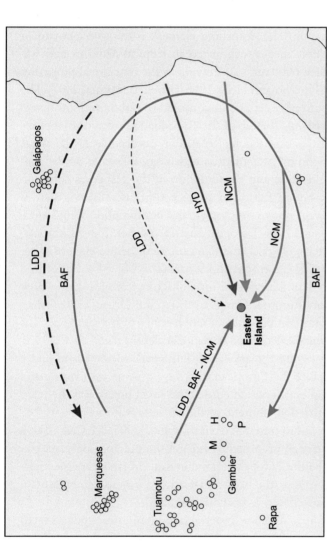

Fig. 2.5 The four existing hypotheses for the arrival of the sweet potato (*Ipomoea batatas*) to Easter Island. According to the Long-Distance Dispersal hypothesis (LDD), sweet potato seeds could have arrived in Polynesian islands by birds, winds or rafting, and then transported by humans to Easter Island (direct LDD to Easter Island seems less likely). The back-and-forth (BAF) hypothesis implies the pre-Columbian arrival of Polynesians to the South American coasts, from where they could have transported the sweet potato to Polynesia and, from there, to Easter Island. The Heyerdahl (HYD) and the Newcomers (NCM) hypotheses propose the direct human transportation of sweet potato from South America. The difference is that the HYD hypothesis contends that the plants were carried by Amerindians before Polynesian settlement (400 CE), whereas the NCM hypothesis postulates that this occurred after Polynesian colonization and does not clarify whether the Amerindians arrived in Easter Island by themselves or were carried by Polynesian in their back-and-forth voyages. H, Henderson Island; M, Mangareva Island; P, Pitcairn Island.

of several species of the genus *Ipomoea*, including *I. batatas*, during prehuman times (Muñoz-Rodríguez et al., 2018). Therefore, the sweet potato could have reached the Pacific, including Easter Island, without the need for human transportation of this species from America. The back-and-forth hypothesis is based on the possibility that sailors from Eastern Polynesia traveled to South America in pre-Columbian times and came back with the sweet potato, thus facilitating its spread over Polynesia (Bourke, 2009). Based on archaeological, ethnobotanical, and linguistic data, it was suggested that the kumara variety of the sweet potato could have dispersed from the Perú/Ecuador coasts to the Marquesas/Tuamotu region (1000–1100 CE), and then rapidly transferred to Hawai'i, Easter Island, and New Zealand (1150–1250 CE) (Yen, 1974; Green, 2000a, 2005) (Fig. 2.6). Remarkably, the Native American name (kumara) has been preserved in Easter Island and New Zealand and their derivatives (koumara, kumala, kumal, uma'a, 'uumara, kuuara, 'uala, umala, and other similar ones) occur across the entire Pacific, from the Micronesian Mariana Islands to Easter Island (Yen, 1974). Historical reports demonstrate that other sweet potato varieties (camote and batata) were transported to the West Pacific by European traders, after the Europe-America contact. Currently, the back-and-forth option is the most accepted hypothesis by archaeologists and ethnobotanists working on this subject.

The back-and-forth hypothesis was originally proposed by Gill (1998) and implies that Polynesians arrived in America prior to Europeans, a view that is gaining momentum (Lawler, 2010). For example, Jones et al. (2011) compiled the material, linguistic, biological, mythological, nautical, chronological, and physical anthropological evidence in support of the pre-Columbian presence of Polynesians in America, and suggested that such contact could have occurred repeatedly between approximately 700 and 1350 CE by three main regions: southern Chile, Ecuador/Perú, and California. According to these authors, Polynesians did not alter the general cultural development of Amerindians but introduced new technologies and domesticates that affected the subsistence practices of local populations. The back-and-forth hypothesis is also supported by recent DNA phylogenies demonstrating that the kumara lineage spread across the Pacific from South American coasts during pre-Columbian times, suggesting that the main responsible for this diffusion were Polynesian sailors in their back-and-forth voyages (Roullier et al., 2013). A decade ago, one of the most convincing evidence of Polynesians traveling to South America was the finding of chicken bones of Polynesian origin as old as

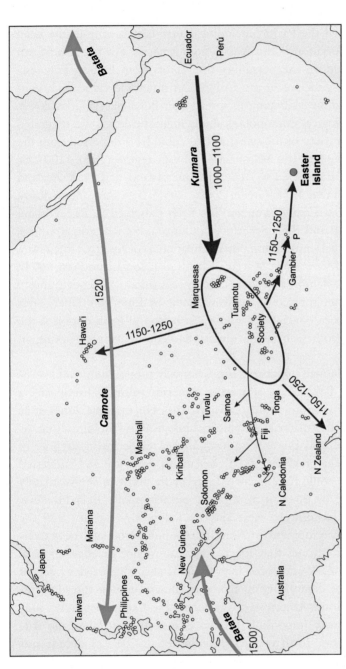

Fig. 2.6 Anthropogenic dispersal of the three varieties of sweet potato: batata *(green)*, camote *(blue)*, and kumara *(red)* across the Pacific, combining archaeological, ethnobotanical, linguistic, historical, and genetic data (Yen, 1974; Green, 2000a, 2005; Roullier et al., 2013). The batata and camote varieties were transported from Mesoamerica to the western Pacific between about 1500 and 1520 CE by Portuguese (batata) and Spanish (camote) traders. The kumara variety is the more widespread across the Pacific and is the variety cultivated in Easter Island. This variety was first transported by Polynesians from South America (1000–1100 CE) and introduced in the Marquesas-Tuamotu-Society region. After establishment in this area, the kumara variety was transferred to Hawaiʻi, New Zealand, and Easter Island between 1150 and 1250 CE. Further dispersals toward the central Pacific occurred in the late 18th-early 19th centuries, facilitated by Europeans.

1300–1400 CE on the Chilean coasts (Storey et al., 2007). The genome of these bones was identical to chicken bones found at prehistoric archaeological sites from Tonga, dated back to 2000–1550 yr BP. However, some authors have attributed this finding to DNA contamination (Thompson et al., 2014). Similarly, some recent studies have been unable to find Amerindian ancestry in a few skeletons from ancient Rapanui people (Fehren-Schmitz et al., 2017), but this seems insufficient to contradict previous results of positive pre-Columbian genetic admixture between the Rapanui and the Amerindians (Thorsby, 2007, 2016).

   The Heyerdahl hypothesis is based on the already mentioned possibility of Native Americans discovering Easter Island long before the Polynesian arrival and carrying the sweet potato with them. To date, this hypothesis lacks archaeological and anthropological support. The newcomers hypothesis may be considered a variant of the Heyerdahl hypothesis, as it postulates that Amerindians could have arrived in Easter Island by themselves when Polynesians were already established. The presence of Native Americans on Easter Island shortly after Polynesian settlement has been supported by DNA analysis of blood from a set of modern Rapanui people without known foreign ancestry. These individuals were predominantly of Polynesian origin (76% of the genome), with an early contribution (8%) from Native Americans dated between 1280 and 1495 CE and a later European contribution (16%) dated to 1850 CE (Moreno-Mayar et al., 2014). However, from these data alone, it is not possible to infer whether the Amerindians arrived on their own (newcomer hypothesis) or were carried by Polynesians after traveling to America (back-and-forth hypothesis) (Thorsby, 2016).

2.3.4 Summary

According to the available archaeological, anthropological, and ethnobotanical evidence, Easter Island seems to have been first settled by Polynesian sailors, probably from the nearby islands (Mangareva, Pitcairn, Henderson), sometime between 800 and 1200 CE. The hypothesis of Heyerdahl, of earlier colonization by Amerindians, seems unsupported but there is evidence for Amerindian presence between about 1300 and 1500 CE (before the European discovery of America), although it is still premature to affirm whether these Native Americans reached Easter Island by themselves or were carried by Polynesians in their back-and-forth voyages between Easter island and South America. Considering the navigation skills of these two cultures, the Polynesian back-and-forth traveling has been considered more plausible. As it has been noted above, Polynesians were sophisticated sailors who colonized the entire Pacific from West to East. One more step reaching South America would have not been surprising. It seems pertinent to stress that youngest settlement dates (1200–1300 CE) rely on the concept of "archaeological colonization" (*sensu* Lipo and Hunt, 2016) and, therefore, the discovery and the first settlement of the island could have occurred earlier and remained still unnoticed due to the scarcity or further removal of the corresponding archaeological evidence (Flenley and Bahn, 2011). As a consequence, it cannot be dismissed that other types of evidence (e.g., paleoecological) is still preserved in other past archives (e.g., lake and swamp sediments), as it occurs in the Azores Islands (Rull et al., 2017; De Boer et al., 2019). We will come back to this issue in the next chapter when discussing the potential usefulness of paleoecology on Easter Island.

2.4 The ancient Rapanui society

Whatever the origin and the time of arrival of the Easter Island discoverers and first settlers, the ancient Rapanui culture was characterized by some developments that are widely accepted and well-documented in the archaeological record, as is the case of the moai cult, the Birdman cult and the island-wide deforestation. Others, for example, the occurrence or not of a prehistoric socioecological collapse, or the maximum population numbers attained by the ancient Rapanui society, are more controversial.

2.4.1 Sociopolitical organization

According to the Rapanui tradition, the first Polynesian colonizers arrived in Anakena commanded by their leader Hotu Matu'a. The island was further subdivided territorially and distributed into six clans (mata) led by the six sons of Hotu Matu'a (Thompson, 1891). Using the same information source, Routledge (1919) reported the existence of ten parental descendant clan groups (Fig. 2.7). According to Métraux (1940), these clans were divided into two main areas, The Tuu or Mata Nui (meaning big clans) and the Hotu Iti or Mata Iti (meaning small clans), roughly corresponding to the western and the eastern parts of the island, respectively. The highly ranked western group (Mata Nui) have its origin in the legacy of Hotu Matu'a to its three older sons, whereas the lower-ranked eastern group (Mata Iti) was composed by the clans formed by the three younger sons (Fig. 2.7). Although some discrepancies exist in some details of this subdivision, the occurrence of a hierarchical district distribution seems obvious (Martinsson-Wallin, 1994; Martinsson-Wallin and Wallin, 2014). It has been

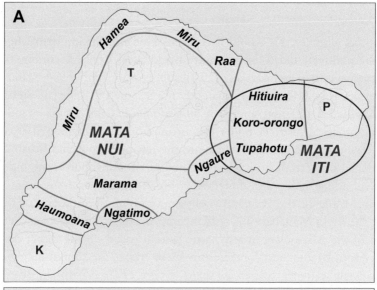

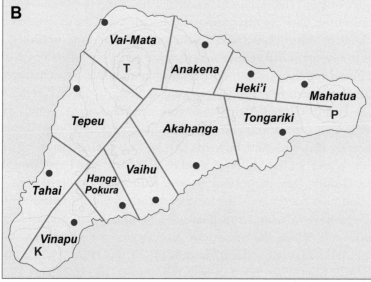

Fig. 2.7 Land distribution in the prehistoric Rapanui society. (A) Clans defined by the descendants of Hotu Matu'a. Note that the inlands and uplands were not assigned to any clan. Clans were grouped into the eastern group (Mata Iti or lower clans) and the western group (Mata Nui or higher clans). (B) Districts defined on the basis of structural details of the largest ahu *(blue dots)*, as indicators group belonging and sociopolitical hierarchy. The interior and coastal uplands are included. T, Terevaka, P, Poike, and K, Kao. *((A) Redrawn from Routledge, K., 1919. The mystery of Easter Island. The Story of an Expedition. Shifton, Praed & Co., London and Métraux, A., 1940. Ethnology of Easter Island. Bishop Museum, Honolulu; (B) Redrawn from Stevenson, C.M., 2002. Territorial divisions on Easter Island in the 16th century: evidence from the distribution of ceremonial architecture, In: Ladefoged, T.N., Graves, M. (Eds.), Pacific landscapes: Archaeological Approaches. Easter Island Foundation, Los Osos, pp. 221–230.)*

discussed whether this would have been the cause for intertribal fighting for the political control of the entire island (Lee, 1992). According to the Routledge (1919) and Métraux (1940) subdivisions, higher and inland areas (Terevaka, Kao, Poike) were apparently not assigned to any clan. Stevenson (2002) used a different approach and subdivided the whole island into 11 hypothetical districts, under the assumption that ceremonial architecture—especially larger ahu, which are the less numerous and situated in a privileged oceanfront position—reflected territorial subdivisions and their corresponding sociopolitical hierarchy (Fig. 2.7).

According to the Rapanui tradition, the highest authority of the island, the Ariki Mau, have his residence in Anakena and was invariably the chief of the Miru clan, in a dynastic system based on the direct descent from Hotu Matu'a and, hence, from Make Make, the supreme deity. The Ariki Mau was seen to possess mana, a type of supernatural power that separated it from the rest and legitimized his leadership role (Robinson and Stevenson, 2017). Other important people were the priests (ivi atua), the chiefs of other clans (ariki paka), and the warriors (matato'a). The access to resources and certain activities was strictly regulated by a strong hierarchical system based on restrictions (tapu), which was characteristic of Polynesian cultures. People working directly in moai carving were considered a privileged class, whose working skills were transmitted from generation to generation, also in a hereditary way, and were forbidden for women (Edwards and Edwards, 2013). This sociopolitical organization was typical of the Middle Period or Ahu Moai, the Expansion Phase or the Golden Age, depending on the chronology utilized (Fig. 2.2). The common feature was the development of the moai industry and cult (Section 2.4.2). After this phase, the moai cult was replaced by the Birdman cult, where the maximum authority of the island was no longer the Ariki Mau, as a direct descendant of the Hotu Matu'a lineage within the Miru clan, but was selected among all clan chiefs after an athletic competition that took place each year during the Austral spring. This cultural shift signified a revolution in the Rapanui life affecting not only political but also the social organization and the religious practices. The Birdman cult has been associated by some archaeologists with the Late Period (Huri Moai) or the Decadent Phase, where natural resources began to be limiting for the prehistoric Rapanui (Fig. 2.2), but other authors believe that the Birdman cult was a postcontact feature (e.g., Robinson and Stevenson, 2017). This is further discussed in the following sections.

2.4.2 The moai cult

There is a general consensus that the moai represented deified ancestors, likely former clan or lineage chiefs, who were worshipped for assuring the fertility of the earth and the sea, as well as for guaranteeing the prosperity of the corresponding clans and lineages. According to Edwards and Edwards (2013), these deities were not considered to be immortal and should be replaced from time to time. The moai were supposed to be under the command of the principal God, Make Make, the only immortal God and creator of the whole humankind. Each moai was unique and have its own name, with the prefix Ariki-. As already stated, most moai were carved on the same quarry, in the tuffs of the Rano Raraku volcano and then transported to their final destination, the ahu. This travel could consist of roughly a kilometer, as is the case of Ahu Tongariki, to 16–18 km (Lipo et al., 2013), in the case of the ahu near Hanaga Roa (Fig. 1.1). How the ancient Rapanui, with their rudimentary technology, may have transported the statues and erect them on their ahu is one of the most popular enigmas of Easter Island and has been subjected to many speculations. As discussed above, the oldest dated ahu is Ahu Nau Nau, in the Anakena beach (Fig. 2.8), whose construction would have initiated between 1200 and 1300 CE (Hunt and Lipo, 2006; Wallin et al., 2010) (Section 2.3.2). Martinsson-Wallin (2001), maintained that the

Fig. 2.8 Ahu Nau Nau, in the Anakena beach. *(Reproduced under CCA 2.0 License (https:// commons.wikimedia.org/wiki/File:Ahu_Nau_Nau_-_Easter_Island_(5956407322).jpg).)*

ceremonial architecture was fully developed slightly earlier, during the 12th century, but the frequent remodelations of the oldest ahu prevent to identify and characterize these initial constructions (see also Cauwe et al., 2007). According to the different chronologies existing, the moai industry would have ceased between 1500 and 1680 CE, depending on the authors (Fig. 2.2). During these 3–4 centuries, the Rapanui built more than 300 ahu and carved almost 1000 moai, which makes an average of roughly 70–100 ahu and 250–300 moai per century or about one ahu and three moai per year. According to Métraux (1940), a modern artist could be able to carve a moai in a few days, but the technology of the Rapanui was much more primitive (Section 1.5). In an attempt to reproduce the ancient moai carving technique, six Rapanui workers hired by Heyerdahl, using basalt tools similar to the prehistoric toki, elapsed 3 days in delineating a profile of a supposed 5-m high statue and estimated that, at the same rate, they would have completed the moai in about 1 year (Heyerdahl, 1989). It may be expected that expert Rapanui carvers would have developed this task in a significantly shorter time and, considering several teams working at the same time, the goal of roughly three moai per year seems feasible.

Regarding transportation, part of the routes by which the moai were moved to reach their corresponding ahu have been mapped (Fig. 1.1); however, the transportation means still remain highly speculative. Several techniques have been proposed and several in situ attempts have carried out to test the diverse proposals. The ancient Rapanui did not know the wheel and lacked large animals such as cattle or horses that could have helped to move the statues. Moreover, the scaffolds, pillars, beams, and ropes necessary to transporting the moai and emplacing them in their ahu would have been impossible to obtain in a totally deforested island. This is why the moai industry has been intimately associated with the phase in which the island was forested and raw materials such as wood and fiber for ropes were available. Heyerdahl was also the first to try to reproduce in situ the transportation and emplacement processes. With the aid of tree trunks, ropes, and stones, a crew of 180 islanders was able to move a moai that was lying on the ground to a nearby ahu and put the statue in a right position. The mystery was declared solved (Heyerdahl, 1989). However, this did not solve the problem of long-distance transportation and the possible damage of statues during it. With the help of the Czech engineer Pavel Pavel, Heyedahl essayed the so-called "refrigerator method," which consisted of making the moai "walk" in a tilting manner, similar to transporting a refrigerator in an upright position, using only ropes and padding. Pavel had already essayed this method in his country and was invited by Heyerdahl to reproduce the technique in Easter

Island, where they succeeded in reproducing the experiment with the help of only 16 workers (Heyerdahl et al., 1989; Pavel, 1990). The handicap of this method was the damage that experienced the base of the moai due to friction. Slightly later, Love (1990) used log rollers and moved a moai replica for a short distance in a few minutes, with the help of 25 men. Sometime before, Mulloy (1974) had proposed that the moai could have been transported in a horizontal position using wooden sledges pushed by many men but he did not essay this method in the field. Van Tilburg and Ralston (2005) used a different sledge method, also in a moai replica, by adding rollers under the sledge structure to minimize friction. More recently, Lipo et al. (2013) used a method similar to the Pavel's refrigerator technique but with some modifications to reduce friction during long-distance transportation (Fig. 2.9). Flenley and Bahn (2011), however, doubt that these modifications would have reduced moai damage at long distances. Not only transporting means but also techniques of emplacing the moai on their ahu have been envisaged by the same researchers, including imaginative methods of capping the statues with their characteristic red scoria hats or pukao (Hixon et al., 2018).

Two observations seem pertinent in regard to the above moai transportation proposals. One is that the use of wood for moai moving would have contributed to deforestation; therefore, the "walking" methods seem to be much more sustainable. However, it is no clear whether this was an argument for ancient Rapanui. Actually, some defenders of the ecocidal hypothesis believe

Fig. 2.9 In situ field experiment of moai moving using a modification of Pavel's refrigerator method (Lipo et al., 2013).

that moai moving was among the main causes of the island's deforestation (e.g., Diamond, 2005). The other observation is that the existing proposals of moai moving may be considered clever attempts to figure out how the ancient Rapanui could have moved and emplaced their statues on the ahu but not much more than that. A number of scholars, including some of the most famous researchers on Easter Island's prehistory, have dedicated tremendous efforts and imagination to this issue by performing field experiments of moai moving. Unfortunately, in the absence of a time machine, this aspect of the moai activity remains untestable. These experiments demonstrate that some techniques are successful in moving a moai (or a moai reproduction) but they do not prove how the ancient Rapanui transported their statues. Moreover, most experiments were carried out on small statues over a flat terrain and for a very short distance, which does not explain how bigger moai were moved, sometimes for up to 18 km and at elevations of 100 and 200 m (Fig. 1.1). In addition, most of the essayed methods used modern cranes in some phase of the field experiments. Other hypotheses have been proposed that cannot be tested by scientific methods, for example, the eventual participation of extraterrestrial beings in the moai carving, transportation, and emplacement (Von Däniken, 1980). According to the Rapanui tradition, the statues "walked" by themselves after the invocations of clan chiefs and priests, using their spiritual power or mana (Thompson, 1891; Métraux, 1940).

The moai industry ceased at some point that is still under discussion (Section 2.2; Fig. 2.2) but some believe that this would have occurred during the mid-17th century (probably between 1625 and 1650 CE) when the island was already deforested (Section 2.4.4) and the Rapanui ran out of the tree trunks needed to transport and install the moai on their ahu (Boersema, 2015a). Recently, however, DiNapoli et al. (2020) considered that the phase of ahu and moai construction continued into the postcontact era. In 1886, William Thompson had the impression that the moai cult ended abruptly, on the basis of personal observations made in situ (Thompson, 1891). For example, a number of unfinished statues remained in the Raraku quarry in different elaboration states, even many that were already finished remained there (Fig. 2.10). Many basalt tools (toki) were scattered as if they were suddenly abandoned to run away. Other finished moai remained unbroken in the transporting routes, suggesting that they had been abandoned during their moving. Moreover, many other moai already in their ahu had been toppled, always upside down and often fragmented, with the head separated from the body. These observations contributed to the idea of a Huri Moai phase, characterized by intertribal violence, which some authors linked to

Fig. 2.10 Examples of finished and unfinished moai still remaining in the Raraku quarry. *(Photo courtesy of O. Margalef.)*

the beginning of prehistoric socioecological demise and a cultural collapse (Section 2.4.5). However, historical documents suggested a different scenario. Neither Roggeveen, in 1722, nor González de Haedo, in 1770, reported a single fallen statue. But in 1774, just 4 years after González's visit, Cook reported many moai toppled and their ahu totally destroyed. The destruction continued for two generations (1770s–1830s) until "…not a single statue remained erect in their ahu" (Fischer, 2005). Therefore, massive and intentional moai toppling did exist but they occurred more than 50 years after European contact, rather than in the Rapanui prehistory (Fig. 2.1). Historians point out that, more than intertribal wars, what contributed to moai and ahu destruction was the fierce rivalry created the Birdman cult, which is explained in the next section.

2.4.3 The Birdman cult

It has been discussed whether the Birdman cult appeared in the prehistory, by the 16th century (Ferdon, 1961; Lee and Liller, 1987), or after European contact (Pollard et al., 2010). Within the first group, some have speculated that the Birdman cult and the moai cult coincided in time and were related in some way (Van Tilburg, 1994; Boersema, 2015a) but the moai cult gradually disappeared and the Birdman cult established as the main island-wide manifestation. The reasons for this cultural shift are not well understood although Robinson and Stevenson (2017) suggested that a territorial restructuring in response to soil nutrient depletion in interior lands, probably due to deforestation, and a long period of dryness would have been involved (Section 2.4.5). Whatever the case, the change from the moai cult to the Birdman cult implied a profound social, political, and religious reorganization in the Rapanui society. The Birdman cult was based on the idea that people not directly descending from the gods could be able to acquire an elevated level of mana not because of his genealogical "merits" (Section 2.4.2) but because Make Make have chosen that particular man as the physical manifestation of himself and, therefore, the maximum authority of the island (Métraux, 1940). The ceremony in which the chosen man was identified and proclaimed as the Birdman (Tangata Manu) was a yearly athletic competition generally performed by the warriors (matato'a) of the different clans. Each year, during the Austral spring, the Rapanui met in the ceremonial village of Orongo from where the athletes should descend the vertical cliffs of the Rano Kao crater by its outer walls to reach the sea and swim until the Motu Nui islet to obtain the first egg of the migratory sooty tern or manutara (*Onychoprion fuscatus*), which was a symbol of fertility (Fig. 2.11). The Birdman and, therefore, the maximum authority of the island was the chief of the clan whose representative was able to return to Orongo with the first egg in intact condition (Routledge, 1919; Métraux, 1940; Englert, 1948).

Orongo and its surroundings constitute one of the most spectacular sites of Easter Island. The ancient ceremonial village of Orongo was formed by approximately 50 oval stone houses simulating inverted canoes, situated on the top of the SW crest of the Kao crater, at about 300 m elevation (Fig. 2.12). Rano Kao and its surroundings are almost devoid of moai and ahu (Fig. 2.1) but are especially rich in rock art. Of the almost 1800 petroglyphs known from around Orongo, nearly 380 depict the Birdman (Fig. 2.13) (Lee, 1992). A detailed archaeological and chronological study of the Orongo area is fundamental to understand the timing and potential causes for the emergence of the Birdman cult, as well as its social and historical impact. However, according to Robinson and Stevenson (2017), the current emphasis on restoration (Section 1.6.1) and—as highlighted by Lipo and Hunt (2016) for Easter Island research, in general (Section 2.2)—the lack of explicit and broad research questions, has led to a still preliminary interpretation of the complex. In spite of this, the work of these authors has already provided relevant insights on the timing of Orongo construction and the temporal trends of the Birdman cult. According to Robinson and Stevenson (2017), former radiocarbon dates of Ferdon (1961) and Orliac and Orliac (1996) for the foundation of Orongo were doubtful, because of their large statistical uncertainty—i.e., more than 4 centuries, in some cases (Fig. 2.14)—and/ or because of the lack of contextual association between the dated material and house construction. Using radiocarbon and obsidian hydration dates on materials related with an extensive petroglyph complex and stone houses, Robinson and Stevenson (2017) concluded that the first activities, likely related with the Birdman cult, took place at Orongo in the early 1600s and,

Fig. 2.11 The travel from Orongo to Motu Nui, in the search for the first egg of the ma-nutara (sooty tern). (A) The Motu Nui islet seen from above the Orongo cliffs. (B) Google Earth image of the pathway *(yellow line)* from Orongo to Motu Nui, including a 300-m descent by the Kao cliffs, a 1.5-km swimming race, a patient waiting (sometimes of several days) for the arrival of the sooty tern, and back. *((A) Photo: V. Rull.)*

after a transitory decline, increased again and attained its maximum in the very late 18th century (Fig. 2.14). According to the authors, the construction of stone houses would have occurred during this late maximum.

Looking at Fig. 2.14 with more detail, the onset of prehistoric activity around Orongo occurred around 1570–1580 CE, with a maximum at 1670 CE and an ensuing minimum at 1740 CE, a couple of decades

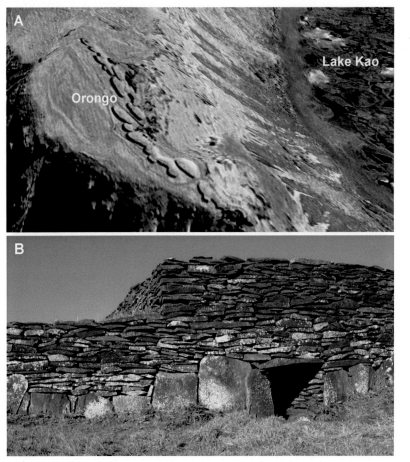

Fig. 2.12 The ceremonial village of Orongo. (A) Google Earth image of the SW crest of the Kao crater showing the elliptic form of the Orongo stone houses, which are currently grass-roofed. (B) Detail of a stone house and its entry. *((B) Photo courtesy of N. Cañellas.)*

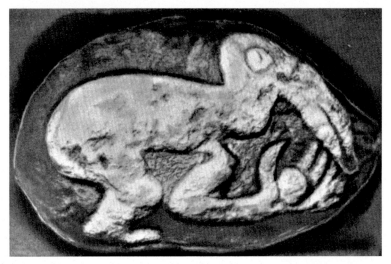

Fig. 2.13 Example of a small petroglyph (40 cm long) representing the Birdman with a manutara egg in the hand. *(Photo courtesy of A. Altman and S. MacLaughlin.)*

after European contact. The absolute maximum activity occurred around 1800 CE, with a further minimum before 1870 CE, a small recovery just before 1880 CE, and the cessation of the activity at 1920–30 CE. It is difficult to compare prehistoric trends with well-dated archaeological evidence of the Birdman cult but events occurred after European contact may be contrasted with historical reports, summarized in Sections 2.1 and 2.4.7. The maximum activity occurred at the beginning of the "Dark Era," characterized by the disruption of Rapanui life by foreign visitors, including

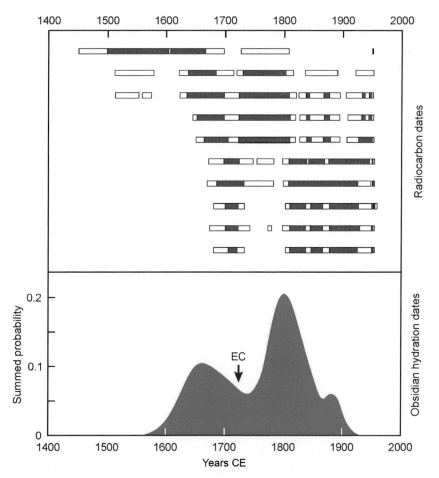

Fig. 2.14 Summary of radiocarbon dates by Orliac and Orliac (1996) (above) and obsidian hydration dates by Stevenson et al. (2013) and Robinson and Stevenson (2017). Radiocarbon dates are expressed as 1α (*brown* bars) and 2α (white bars) intervals. EC, European contact. *(Composed and redrawn from Robinson, T., Stevenson, C. M., 2017. The cult of the Birdman: religious change at 'Orongo, Rapa Nui (Easter Island). J. Pacif. Archaeol. 8, 88–102.)*

frequent violence episodes (Boersema, 2015a). According to Robinson and Stevenson (2017), this could have influenced the ritual behavior of Easter Islanders although these authors did not mention the specific process by which this might have occurred. Fischer (2005) pointed out that the rivalry among the different clans competing for the Birdman exacerbated after European contact and degenerated into an open battle for destroying their competitor's mana, whose most conspicuous manifestations were their moai. The battle was especially intense between the eastern and western clan groups (Section 2.4.1). The Birdman cult was prohibited by the catholic missionaries in 1862, at the beginning of the postcontact genocide (Section 2.4.7), but the contest continued until 1879, albeit purely as a sports competition (Routledge, 1919).

2.4.4 Deforestation

At present, there is little doubt that Easter Island was deforested by the Rapanui before European contact (but see Hunter-Anderson, 1998). Most researchers are rather vague regarding the percentage of the island that was forested but the general impression they give is that forests covered the whole island, or almost. Mieth and Bork (2015, 2017) did more precise estimations and suggested that 80% of the island would have been covered by dense forests containing almost 20 million palm trees (Fig. 2.15). Deforestation causes and timing are still subjects of active debate. Regarding causes, the most accepted view is that the ancient Rapanui society was mainly responsible for forest clearing, either in a direct or indirect manner. Straightforward human deforestation was deduced by palynological evidence of palm forest removal and fire increase coinciding with human arrival to the island (Flenley and King, 1984; Flenley et al., 1991) (Chapter 4). It was also suggested that rats carried by the Polynesian settlers could have contributed to the deforestation. This is based on the finding of fruit endocarps likely belonging to the extinct palm (Section 1.4.3), showing evident signals of rat gnawing, which suggested that fruit eating by rats could have contributed to deforestation by preventing forest regeneration (Dransfield et al., 1984). According to Hunt (2006, 2007), this would have been the major cause of deforestation. However, Mieth and Bork (2010) pointed out that the fruits used as evidence were fully exposed in caves and, therefore, were accessible to rats for long periods before their finding. When these authors analyzed endocarps buried in situ within charcoal layers, they found that only 10% of them having gnaw marks and, furthermore, forests had been able to regenerate after the first clearing (Section 6.1.3). These authors concluded that rats did not hinder forest regeneration and, hence, were not

Fig. 2.15 Comparison between a hypothetical reconstruction of former Easter Island's palm forests (A) and the present landscape (B) in the Poike peninsula. *(Images courtesy of A. Mieth.)*

mainly responsible for island-wide forest clearing (Section 6.1.3). It has also been speculated that some diseases or parasites common in several palm species could have participated in forest clearing without human intervention (Ingersoll and Ingersoll, 2017), but no evidence has been provided thus far. Another speculative hypothesis is that the ancient Rapanui used the sap of the palm trunks for different purposes and this would have also contributed to deforestation (Bork and Mieth, 2003). This proposal is based on the significant volume of sap in the trunks of *Jubaea chilensis* from continental Chile and its well-documented historical use.

Deforestation timing and chronology will be discussed after analyzing paleoecological evidence in detail but it can be advanced that the first estimates, on which most archaeological discussions are based, vary between 800 and 1250 CE for the beginning of forest clearing, and 1450–1650 CE for the total forest removal (Flenley and Bahn, 2003; Hunt, 2006, 2007; Mieth and Bork, 2010). Some authors argue that deforestation was not abrupt but started at 800 CE as a small-scale practice and intensified by 1250–1350 CE (e.g., Mieth and Bork, 2015). The defenders of the ecocidal hypothesis consider that the assumed prehistoric collapse of the Rapanui culture is intimately linked to deforestation (Section 2.4.5). However, increasing evidence supports that forest clearing and sociocultural demise were two separate events, both in time and causes. Archaeological evidence in this sense will be summarized in the next section and paleoecological evidence will be extensively discussed in Chapters 4–6.

2.4.5 Collapse or resilience?

La Pérouse (1797) already speculated that Easter Island might have been forested and that the ancient Rapanui would have cut down all the trees making the island inhabitable. But was William Mulloy (Fig. 2.16) who, in a popular paper (Mulloy, 1974; quoted expressions are literal), explicitly considered the island as a "miniature universe" that was "formerly wooded" but was progressively cleared to make more room for agriculture in the way toward a prosperous, numerous, and diverse society that coincided with the maximum development of the moai industry. This flourishing, and the confidence on the protection of their deified ancestors represented in the moai, probably created in the islanders the "solid assurance that their success was permanent." However, "disaster hovered and it was not precipitated by enemies from beyond the sea." Indeed, "45 square miles was a finite environment" and although "food-producing potential was probably never completely exhausted, its limits may have been approached." As a consequence, "the hitherto efficient economic equilibrium disintegrated," wars appeared, food production was seriously compromised and the moai industry ceased. As the "catastrophe" progressed, "land lay fallow and the unused plantations developed a sparse short grass cover quite different from the original protective bush," which caused the "formerly retained moisture sank quickly into the soil and escaped into the sea." This led to a scanty and culturally

Fig. 2.16 William Mulloy typing field notes during the Norwegian expedition led by Thor Heyerdahl in 1955. *(Reproduced under license CCA-SA 3.0 (https://en.wikipedia.org/wiki/File:Mulloy_typewriter.jpg).)*

degraded society that the first Europeans encountered on the island. But the degradation continued and further "slave raids and their sequelae wrote the epilogue of a spectacular culture that would appear to have been unable to cope with a population too numerous to maintain the social relationships by which it had adapted to its tiny environment." Mulloy (1974) ended by suggesting that "in a typically human fashion few lessons were learned from this devastating past" and that "this unfortunate history could repeat itself."

Mulloy's (1974) paper seem to have inspired some researchers to seek for evidence of full island's deforestation and self-induced social and cultural catastrophe, which was later known as the ecocidal theory (Diamond, 2005) and guided further Easter Island research. A decade later, the first palynological analyses performed on rano sediments (Flenley and King, 1984; Flenley et al., 1991) seamed to strongly support Mulloy's ideas on deforestation (see Chapter 4 for details) and the ecocidal explanation became firmly rooted in many scientific and nonscientific media. As discussed in Section 2.2, several authors situated the prehistoric degradation of the ancient Rapanui society and its final collapse at different ages, ranging from 1500 to 1680 CE (Fig. 2.2). This phase is commonly called Huri Moai of Decadent Phase and has been typified by a cluster of events including the shortage or the exhaustion of natural resources by overexploitation, a significant population decline (Section 2.4.6), the abandonment of the moai cult and the intentional toppling of these statues along with the destruction of their ahu, the shift from cremation to burial practices using the ahu remains (Shaw, 1998), the change of lifestyle habits from outdoor dwellings and activities to a more subterraneous life within conditioned lava-tube caves (Ciszewski et al., 2009), the onset of social conflicts, wars and cannibalism, and the development of the Birdman cult, intimately associated to the rock art, mainly by petroglyph carving (reviews in Bahn and Flenley, 1992; Flenley and Bahn, 2003; Diamond, 2005; Bahn, 2015). Even the statues and their pukao were carved with petroglyphs, which has been suggested to be a sacrilege for the moai cult (Lee et al., 2017). The archaeological support for sustaining this dramatic cultural shift is abundant—moai and ahu destruction, ample evidence of cave occupation, abundant skeletons with assumedly intentional injuries—but one of the most characteristic objects are the small (5–10 cm, in average) obsidian flaked tools (mata'a), which were interpreted as warfare elements, usually spear points (Fig. 2.17).

However, a number of archaeologists, despite recognizing that the island was totally deforested during the prehistory, did not find convincing evidence of a cultural and demographic crisis. These scholars argued that the ancient Rapanui were resilient to forest clearance and remained as a healthful society until European contact, after which they collapsed due to slave trading and the introduction of alien epidemic diseases such as smallpox, syphilis, and tuberculosis (Section 2.4.7). Therefore, the Rapanui cultural collapse would have been a genocide, rather than an ecocide (Rainbird, 2002; Peiser, 2005; Hunt, 2006, 2007). Further archaeological evidence supported the healthy continuity of the Rapanui society until European arrival. Indeed, widespread land-use evidence and radiocarbon dates from archaeological sites across the island suggest that the ancient Rapanui developed sustainable agricultural practices and did not experience a social crisis prior to European arrival.

The ancient Rapanui did not practice extensive cultivation or widespread irrigation; crops were restricted to meter-scale garden features called manavai (Fig. 2.18), whose water supply was heavily dependent on rainfall (Hunt and Lipo, 2011; Puleston et al., 2017). To minimize evaporation and erosion, the manavai were protected from wind and surficial runoff by rock walls up to 2 m high, and the soils inside were covered by rocks to preserve heat, moisture, and nutrients, a practice called lithic mulching (Stevenson et al., 1999; Wozniak, 1999, 2001; Gossen and Stevenson, 2005). Chemical analyses of agricultural soils suggested that they were more nutrient-rich than nongardened

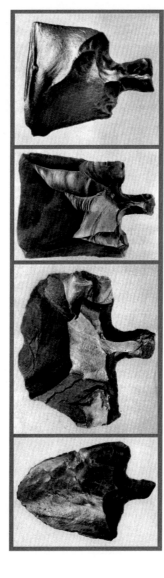

Fig. 2.17 Some examples of obsidian mata'a from several sources. *(Photo courtesy of A. Altman and S. McLaughlin.)*

Fig. 2.18 A modern manavai in the Botanical Garden of Hanga Roa with a cultivated toromiro (*Sophora toromiro*) plant inside. *(Reproduced under license CCA-SA 3.0 (https://commons.wikimedia.org/wiki/File:ToromiroRapanui.jpg).)*

soils, likely due to mulching-reduced leaching or active release of household organic matter (Hunt and Lipo, 2011; Ladefoged et al., 2005, 2010). Based on extensive field surveys, the first estimations yielded a surface of $76 \, km^2$ of mulching area (46% of the island surface), during a period between about 1460 and 1860 CE (Bork et al., 2004). Using satellite imagery, Hunt and Lipo (2011) identified more than 2500 manavai-like cultivation structures, for a total of $16.5 \, km^2$, which represents 10% of the island's surface. Later, these estimates were increased to 13% of the coastal zone and 19% of the total island's surface (Ladefoged et al., 2013). According to Mieth and Bork (2005), during the first 500 years of human occupation, cultivation was carried out along the coasts and 1–2 km inland, and the main volcanic peaks (Kao, Pike, and Terevaka) were likely forested; however, after approximately 1500 CE, these forests were also replaced by manavai structures.

Recent investigations have suggested that these agricultural practices continued, with no significant decline, until European contact (Mulrooney, 2013; Mieth and Bork, 2015, 2017; Stevenson et al., 2015; Jarman et al., 2017; Wozniak, 2017). Robinson and Stevenson (2017) suggested that some reduction of arable land could have existed due to soil erosion after deforestation but this would have led to spatial reorganizations of agricultural practices leading to cultural continuity, rather than a population collapse. These spatial reorganizations were common in other Pacific archipelagos, where inland migration have been documented (Nunn, 2007). In addition, Polet (2015) and Polet and Bocherens (2016), using dental microwear and carbon and nitrogen isotope analysis of skeletons dating from the 17th to the 19th centuries, concluded that the Rapanui of those times have a well-balanced diet and did not find evidence for severe malnutrition and cannibalism, as proposed by the collapse hypothesis. The diet of the prehistoric Rapanui included terrestrial and marine components in variable proportions, according to the living site and the social status of the individuals (Commendador et al., 2013; Jarman et al., 2017). In addition to the nutrients supplied by cultivated plants mentioned in Section 1.4.3 and marine algae, animal protein was provided by rats, chickens, terrestrial and marine birds, eels, octopuses, shellfish, turtles, a variety of reef and offshore fishes, and marine mammals such as dolphins (Commendador et al., 2019). After detailed reviews of the available evidence, several archaeologists concluded that, despite its popularity, the idea of a prehistoric population collapse is not supported by archaeological and historical data (e.g., Hunt, 2007; Love, 2007; Mulrooney et al., 2007). In the words of Love (2007): "archaeological evidence for how the hypothesized prehistoric cultural collapse of Easter Island's society came about has virtually nonexistent, but rather extensively assumed by much of the scientific community, and mostly by the press."

In addition to this agricultural prosperity, several archaeological and anthropological evidence previously considered as indicative of social violence by the defenders of the ecocidal hypothesis was reinterpreted to being more characteristic of a peaceful society. The most relevant features in this sense are the mata'a and the injured skeletons. For example, Church (1998) and Church and Ellis (1996), after in situ studies of mata'a found both in caves and outdoor sites, considered that these objects were used for peaceful purposes such as food preparation, gear fabrication and maintenance, sawing, cutting fresh plants, cutting meat, cutting and scrapping chicken bones, scrapping fish and whittling softwood, among others, and that it was certain specialization across sites (see also Ayres et al., 2000). A detailed morphometric analysis of more than 400 mata'a, led Lipo et al. (2016) to conclude that these objects were not used as lethal weapons but for peaceful activities such as cultivation or ritual scarification. More recently, Stevenson and Williams (2018) performed obsidian hydration dating (Stevenson and Novak, 1988; Stevenson et al., 2013) on more than 60 mata'a and observed that their use peaked in the 16th century and then declined. After an experimental study of tool breakage patterns, these authors concluded that the mata'a would have been used for forest clearance. Regarding cranial injuries observed in skeletons from ancient Rapanui, Owsley et al. (2016) pointed out that most of the scars and fractures were compatible with episodic and localized violence events, rather than with island-wide warfare.

Recently, Bahn (2015)—who, together with Flenley, has been one of the main defenders of the ecocidal view—emphasized that, in the third edition of their famous book on the subject (Bahn and Flenley, 1992), they decided to replace the term "collapse" by "decline," which is more accurate and less dramatic. However, Bahn (2015) maintained the existence of starvation and violence caused by total deforestation and massive soil erosion and, albeit he admitted that the Rapanui adapted to the new circumstances and developed a sustainable agriculture system based on lithic mulching, he also emphasized that the islanders were not in good condition when Europeans reached the island, as compared to former times. Finally, Bahn (2015) also accepted that the true collapse of the Rapanui culture occurred as a consequence of European contact. Boersema (2015b, 2017) disagreed with the idea of a precarious Rapanui society at the time of European contact. According to this author, the bad impression caused by the Rapanui society to the first Europeans may have largely been influenced by the accounts of Cook's expedition. The first explorers, Roggeveen and González, had described a rather healthful Rapanui society, but Cook and his crew—probably frustrated by not finding a situation similar to other Pacific islands, where to take fresh supplies and water on board—described the island as disastrous, in terms of both natural resources and cultural development (Boersema, 2015b, 2017).

2.4.6 Demography

As occurs with other Easter Island issues, reconstructing the population dynamics of the ancient Rapanui society is a highly speculative task (Anderson, 2002). The more reliable numbers are those provided by the first European explorers but the prehistoric population trends have had to be inferred using modeling methods with initial assumptions based on either ecocidal or resilient premises (Merico, 2017), or on estimates based on residential features, agricultural yields and population densities of similar islands (Puleston et al., 2017). The "magic number" pursued by most authors has been the peak Rapanui population at the time of the full development of this culture. The first European explorers estimated a population of up to 3000 inhabitants (Fischer, 2005; Boersema, 2017), a number that was considered insufficient to produce the spectacular sociocultural development expressed in the archaeological record, especially the carving, moving, and emplacement of nearly 1000 multiton moai and the construction of their respective ahu (Section 1.5). This view implicitly involved the idea of a significant population decline, likely linked to a cultural collapse (but see Puleston et al., 2018).

Contrarily, Hunt and Lipo (2009) and Lipo et al. (2018), using noncollapse assumptions, contended that there is few archaeological evidence that the population was ever much larger than the 3000 individuals, as seen by the first European explorers. Estimates of the prehistoric Rapanui population peak using archaeological considerations, agricultural production, comparison with the population density of other Polynesian islands and modeling are extremely variable, ranging from 3000 to 25,000 people, with an average of approximately 10,000 (Table 2.1).

Regarding mathematical modeling techniques, Merico (2017), classified the relatively few demographic models existing for Easter Island into two main categories, those that focus on socioeconomic and those emphasizing ecological aspects. The first group commonly use Mathusian methods of population growth limited by natural resources (Malthus, 1798), whereas the second approach is based on Lotka-Volterra predator-prey methods (Lotka, 1925; Volterra, 1926), in which humans are the

Table 2.1 Different population estimates for the prehistoric Rapanui demographic peak.

Estimated population	Method/evidence	References
3000–4000	Estimated population density	Métraux (1940)
3000–4000	Estimated growth rates	Hunt (2007)
3000–4000	Modeling (ecological)	Basener et al. (2008)
4000	Modeling (genocidal assumptions)	Brandt and Merico (2015)
7000	Modeling (slow demise assumptions)	Brandt and Merico (2015)
7000–9000	Archaeology (several sources)	Van Tilburg (1994)
10,000	Modeling (socioeconomic)	Brander and Taylor (1998)
10,000	Modeling (ecological)	Basener and Ross (2005)
10,000	Modeling (socioeconomic)	Good and Reuveny (2006)
10,000	Modeling (socioeconomic)	Roman et al. (2016)
10,000	Forest resources, charcoal, soil erosion	Flenley and Bahn (2003)
12,000	N.A.	Fischer (2005)
15,000	Modeling (socioeconomic)	Mahon (1998)
15,000	Other's archaeological estimates	Diamond (2005)
15,000	Estimated population density	Fagan (2008)
15,000	Modeling (ecocide assumptions)	Brandt and Merico (2015)
17,500	Agriculture carrying capacity	Puleston et al. (2017)
25,000	N.A.	Rallu (2007)

N.A., not ascertained.

predators and natural resources are the prey. Among those models based on socioeconomic parameters, the first was that of Brander and Taylor (1998), who estimated a maximum population of about 10,000 individuals by 1250 CE and a decline to <4000 by 1850 CE. This model has been refined with time by the progressive addition of economic parameters such as minimum resource consumption per capita, resource variability, thresholds in the resource stock, adaptation by technical progress or economical stratification. All these models predicted that a collapse was inevitable unless population growth was sufficiently slow or resource renewal was sufficiently fast (Merico, 2017). Another type of socioeconomic models considered social learning and foresight, that is, care about the future, by introducing property rights, optimal resource management, and centralized planning (Good and Reuveny, 2006). However, the results obtained differ little from those of Brander and Taylor (1998) model and its further modifications.

Models that use ecological features in their premises were first developed by Basener and Ross (2005) and Basener et al. (2008), who introduced parameters such as the relationship between harvesting and resource renewal rates and palm fruit consumption by rats (Section 2.4.4). The results are quite distinct if rats are introduced or not in the models. If fruit consumption by rats is not considered, the human population attains a maximum of 10,000 individuals and then rapidly crashes (Basener and Ross, 2005). When rats are considered, the human population only crashes if rats are decisive deforestation agents; otherwise, human-rat coexistence is possible and human population stabilizes at 3000–4000 individuals with no further collapse (Basener et al., 2008). Recently, Brandt and Merico (2015) developed a modified version of these models including the introduction of epidemic disease after European contact (Section 2.4.5). This model was used to test three scenarios: the prehistoric ecocide, the postcontact genocide, and a combination of both. Using ecocidal parameters, Brandt and Merico (2015) obtained a population acme of about 15,000 inhabitants by 1600 CE with a sudden decrease to 3000 during the European contact. Under the genocidal assumptions, the population attained a maximum of 4000 inhabitants by 1400 CE, which was maintained until 1750 CE, when rapidly declined to <1000 by 1850. Combining the ecocidal and the genocidal scenarios, the model showed a maximum of almost 7000 people by 1400 CE with an ensuing slow decrease to 3000 by the European arrival, and a further decline to 1500 by 1850 (Fig. 2.19).

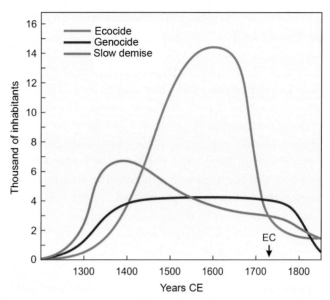

Fig. 2.19 Comparison of the three outputs of a recent demographic model focused on ecological aspects, under three different scenarios (ecocide, genocide, and a combination of both). *(Composed and redrawn from Brandt, G., Merico, A., 2015. The slow demise of Easter Island: insights from a model investigation. Front. Ecol. Evol. 3, 13.)*

2.4.7 The genocide

Although this section goes beyond the main temporal scope of the book, it is necessary to highlight that the Rapanui genocide really occurred, albeit it took place after European contact, and is very well documented historically. This genocide has two main components, one is population demise and the other is the acculturation of the few remaining Rapanui people. The arrival of the first European expeditions dramatically changed the prehistoric Rapanui society, which entered abruptly into history, but in the history of others, rather than in its own. The following summary is based on Fischer (2005) and Boersema (2015a). The frequency of visits to Easter Island between the late 18th century and the mid-19th century increased spectacularly and profoundly disturbed the Rapanui society with epidemic diseases and abuses of any type until the point that Easter Islanders developed an intense xenophobia and any landing attempt of foreign ships were received with stone throwing. In the second half of the 19th century, however, arrived the coup de grace that represented the beginning of the end of the ancient Rapanui culture. By that time, slave raids organized from Perú, where slavery was still legal, were common in Polynesia but they involved a few people. However, this very productive commerce intensified in 1862 and reached Easter Island.

The worst episode occurred with the arrival of the Spanish buccaneer Captain Maristany—known by the Rapanui as Marutani (Amorós, 1992)—who violently kidnapped almost 350 Rapanui to be sold to Perú for working on guano extraction and in domestic labors. This was followed by more expeditions that totalized more than 1400 slaves in 6 months, roughly a third of Easter Island's population (4000) by the time. In 1863, human traffic was prohibited in Perú, the repatriation of Easter Islanders, among others, began and 470 Rapanui were embarked to return to their original home. However, these "fortunate" people had already contracted, in Perú, infectious diseases unknown to them, such as syphilis, smallpox, or tuberculosis, and only 15 of these Rapanui arrived alive to the island. Smallpox and tuberculosis spread over the island and, a decade later (1877), only 110 Easter Islanders remained alive (Fig. 2.20) (Pinart, 1878). Meanwhile, the acculturation was in progress and, in 1868, the Franco-Belgian missionaries, led by Eugène Eyraud, have already Christianized the entire island's population, which was of 800 Rapanui. Eyraud died the same year of tuberculosis. In summary, a real collapse of the Rapanui culture, in the form

Fig. 2.20 The French philologist Alfonse Pinart (left, hat in hand) visited Easter Island in Easter Sunday 1877 and reported that only 110 Rapanui were left after the smallpox and tuberculosis epidemics (Pinart, 1878). *(Image courtesy of A. Altman and S. MacLaughlin.)*

of a genocide, occurred between 1862 and 1877, when the population almost disappeared and the few remaining people had been acculturated (Fig. 2.1).

Ingersoll et al. (2017) summarize the debate about the collapse of the Rapanui culture as follows:

> *Rapa Nui ecocide is a convenient untruth promoted by environmentalists looking for an apocalyptic* **Lorax** *tale that appeals to the popular press. The Rapanui were enslaved, extradited, battered by disease, suffered rendition, relieved of their land, and then, we have not mentioned this yet, effectively interned in the village of Hanga Roa until the 1960s. Scholars and Western nations: stop blaming the victim and take responsibility for what really happened. Credit the Rapanui for unparalleled resilience in the face of adversity. Today, Rapanui culture is alive, well, and vibrant, against all odds.*

2.4.8 Summary

Some hints on the prehistory of Easter Island may be extracted from this short review. First, it should be emphasized that a tentative, although rather simple, chronology can be suggested on the basis of the evidence discussed in Sections 2.4.1–2.4.7, avoiding references to oral tradition (Fig. 2.21). This tentative chronology may be considered a succession of events in time with no strict boundary dates and has been attempted only as a trial to set a preliminary framework for further discussions. This preliminary framework will be revisited later when all paleoecological evidence will be disclosed (Chapter 7). For now, a general observation is that the shift from the moai cult to the Birdman cult seems to have coincided with the latest estimates for island–wide deforestation (1600–1650 CE). It is also noteworthy that the moai toppling and the demographic and cultural collapse (i.e., the genocide) of the Rapanui society were not prehistoric but postcontact events that did not coincide with the onset of the Birdman cult, as suggested by the ecocidal hypothesis and a number of chronologies reviewed above (Fig. 2.2). Regarding population numbers, only the historical estimates made by the first Europeans seem to be reliable, as quantitative inferences for the time of flourishment of the ancient Rapanui culture, likely occurred during the moai cult phase, are deeply discordant and often rather speculative or based on biased assumptions toward either one or another hypothesis (i.e., ecocide, resilience, or genocide). Therefore, neither the postdeforestation population crash proposed by the ecocidal hypothesis nor the lower-level population stability suggested by the resilience option may be properly tested with the available data.

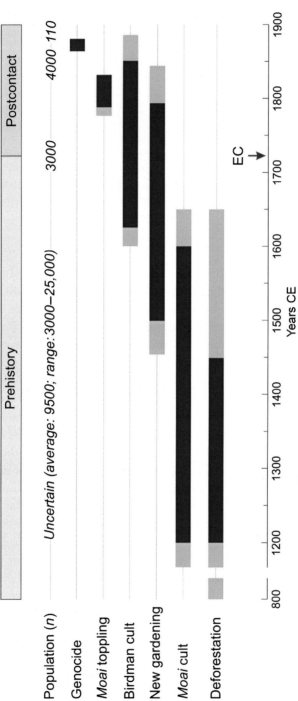

Fig. 2.21 Tentative chronological summary of the different processes discussed in Sections 2.4.1–2.4.7. Transition phases and time intervals under discussion are indicated in *light blue*. Population numbers according to Table 2.1. EC, European Contact.

In summary, according to the archaeological, anthropological, and historical evidence reviewed in this chapter, the prehistory of Easter Island could be tentatively subdivided into two main phases, the first characterized by the deforestation (800–1200 to 1450–1650 CE) and the moai cult (1200 to 1600–1650 CE), and the second characterized by the incipient Birdman cult (1600–1722 CE). The maximum development of the Birdman cult (1800–1850 CE), the toppling of the moai (1770s–1830s), and the genocide (1860s–1870s) seem to have actually been postcontact features (Fig. 2.21).

2.5 A narrative of human determinism

Regardless of the preferred or better-supported hypotheses on the prehistory of Easter Island, all the approaches reviewed in this chapter have a common feature: all that has happened seems to have been a consequence of human activities. This is not surprising, as most inferences done so far have been based on archaeological, anthropological, ethnographical, and historical evidence. Does this mean that environmental change, especially eventual climatic shifts, have had no influence on ecological and cultural transformations? This question cannot be answered without examining the available evidence for climate change and the potential links between climatic, ecological, and human sequence of events and their respective interactions. To develop such an integrated approach, care must be taken not to fall into unwarranted deterministic attitudes. The approaches reviewed in this chapter are in line with human determinism—not to be confused with cultural determinism—that considers cultural developments as a consequence of exclusively human actions. Contrarily, environmental determinism assumes that most cultural changes are driven by environmental shifts that overcome the human reaction capacity. See Rull (2020c) for a more detailed discussion on these approaches and illustrative examples. This book uses a holistic approach, free of a priori determinisms. However, for this to be accomplished, it is necessary to explain in detail the existing paleoecological evidence of past environmental change, which has been traditionally ignored or explicitly neglected by many researchers. The next chapters will concentrate on paleoenvironmental change in a trial to balance the situation but the reader should not worry since the final chapters will address the Easter Island's prehistory under an integrated climatic-ecological-cultural integrative framework.

CHAPTER 3

Introduction to Easter Island's paleoecology: Why, where, and how?

Contents

3.1	Why study paleoecology?	90
	3.1.1 General considerations	90
	3.1.2 The case of Easter Island	93
3.2	Initial proposals of prehistoric climate change on Easter Island	95
3.3	Coring sites	100
	3.3.1 Rano Aroi	100
	3.3.2 Rano Kao	101
	3.3.3 Rano Raraku	102
3.4	Cores retrieved and main proxies studied	103
3.5	Chronology and sedimentary patterns	110
3.6	Other paleoecological archives	116
Appendix 3.1		117

This chapter begins with a section that highlights the usefulness of paleo-ecological research, in general, and its application to Easter Island, in particular, before entering into the more methodological and descriptive parts. The initial section also examines the first speculations about the potential role of climate change on Easter Island's prehistory, before paleoecological evidence began to shed light on the local climatic history of the island. The rest of the chapter is a descriptive account of all paleoecological archives studied to date on Easter Island, including all cores retrieved and all the paleoclimatic and paleoecological proxies analyzed in them. This is essentially a methodological chapter to show where and how the paleoecological results and interpretations disclosed in Chapters 4–6 have been obtained. This chapter also includes a database of all radiocarbon ages obtained to date in the cores listed. Although the book is focused on Easter Island's

Paleoecological Research on Easter Island
https://doi.org/10.1016/B978-0-12-822727-5.00003-9

prehistory, this chapter encompasses the whole temporal range of paleoecological studies developed on the island's archives, ranging from about 70,000 cal yr BP to the present.

3.1 Why study paleoecology?

3.1.1 General considerations

Paleoecology has been defined as "the ecology of the past" (Birks and Birks, 1980) or, more precisely, "the branch of ecology that studies (the) past (of) ecological systems and their trends in time using fossils and other proxies" (Rull, 2010). Almost 40 years ago, the North American paleoecologist Andrew Hill asked: "why study paleoecology?" and concluded that paleoecologists should focus on solving interesting problems, rather than on gathering more and more data just because it is possible to do it (Hill, 1981). Noteworthy, this message is very similar to that of Lipo and Hunt (2016) and Robinson and Stevenson (2017) addressed to archaeologists working on Easter Island (Sections 2.1 and 2.4.3). Regarding the so-called interesting problems to be addressed, many paleoecologists—especially those working on the Quaternary—have concentrated on aspects such as the ecological dynamics over long-term timescales, the ecological legacies from human activity and environmental change, the drivers and causal mechanisms for current landscape patterns, the extent of natural environmental variability, or the responses of organisms to this variability and their usefulness for predicting future biotic changes (Birks, 2008). In addition, paleoecology is useful to inform conservation and restoration practices in a wide range of spatial and temporal scales (Vegas-Vilarrúbia et al., 2011; Willis et al., 2010). A review of other ecological, evolutionary, and biogeographical problems that can be addressed using Quaternary paleoecology can be found in Rull (2020c), but those highlighted above seem enough to realize that the prehistory of Easter Island is, beyond all doubt, among the interesting subjects that paleoecology is able to address. There is no need to say that paleoecology cannot furnish evidence on all the engimas of Easter Island. For example, paleoecology alone cannot provide insights on cultural matters such as the meaning of the moai and the religious practices associated to them, or on the means of transportation of these statues from the quarry to the ahu and its final emplacement. However, paleoecology is able to provide information about the timing of settlement, the potential influence of climate, and other environmental drivers on ecosystems and their eventual cultural influence, as well as on the ecological impact of human practices. This information, combined with archaeological, anthropological, and ethnographical evidence, is useful to move toward an integrated view of Easter Island's prehistory. The usefulness of paleoecological research for the understanding of settlement features and further socioecological developments has been demonstrated in many islands worldwide, including those of the Pacific Ocean (e.g., McGlone, 1983; Kirch and Ellison, 1994; Flenley, 2007; Nunn 2007; Prebble and Dowe, 2008; Prebble and Wilmshurst, 2009; and literature therein), but its full potential is yet to be exploited on Easter Island.

As demonstrated in the former example of the Atlantic Azores Islands (Section 2.3.1), an important part of the evidence of human activities may be absent in the archaeological record, but may be preserved in lake sediments, which has the comparative advantages of being fairly continuous and well preserved. There are many other examples of paleoecological records of early human presence before archaeological colonization. In the present context, the most usual sedimentary proxy for human presence is the pollen of anthropochore plants, which has been decisive for documenting prehistoric settlement events and migrations (Faegri et al., 1989; Goudie, 2006; Roberts, 2014). Recently, this field of research has benefitted from the amazing development of molecular analytical techniques, which allows identification of molecular biomarkers of human presence, such as DNA or specific fecal lipids, in lake sediments (e.g., Bull et al., 2002; D'Anjou et al., 2012; Hofreiter et al., 2012; Rawlence et al., 2014; Parducci et al., 2017). A continuous paleoecological record can be of much help to complement archaeological evidence, whose discontinuous nature is evident. In addition, the archaeological record that is exposed, as is the case of most of the objects of the "Easter Island outdoor museum", is more susceptible to deterioration by vandalization and exposure to atmospheric agents, total or partial removal by heritage piracy, or intentional destruction by social conflicts or by further colonizers (Sections 1.5 and 1.6).

It is very illustrative to compare the exposed archaeological record with the terrestrial glacial record. Until less than half a century ago, the Quaternary glacial history was entirely based on their terrestrial manifestations, especially erosional and depositional landforms such as cirques, U-shaped valleys, hanging valleys, moraines, fluvioglacial sediments, and outwash fans, among others (Fig. 3.1). Based on these exposed features, four main glaciations were found to have occurred during the last 600,000 years. In central Europe, these glaciations were named as Günz (592–543 kyr BP), Mindel (478–429 kyr BP), Riss (236–183 kyr BP), and Würm (115–11.5 kyr BP) (Penk and Brückner, 1901–1909; Köppen and Wegener, 1924). Their North American counterparts were the Nebraskan, Kansan, Illinonian, and Wisconsin glaciations, respectively (Flint, 1971). The situation, however,

Fig. 3.1 Panoramic view of a typical glacial landscape, in the highest part of the SW Europe Pyrenean range (up to 3400 m elevation), with a set of characteristic glacial landforms mentioned in the text. *(Photo: V. Rull.)*

underwent a radical change after the 1960s with the study of marine sediments obtained by deep ocean floor drilling. Using paleoclimatic proxies such as oxygen isotopes, it was possible to establish the sequence of global glacial-interglacial cycles that took place since the beginning of the Quaternary (2.6 Ma) (Fig. 3.2). In this way, more than 40 glaciations were identified during the Quaternary, which had a periodicity of 41 kyr to 100 kyr (Raymo, 1994). Therefore, the current terrestrial glacial record of central Europe is limited to four of the last glaciations occurred and the rest are missing. The erosional strength of the huge ice sheets and glaciers of these last four glaciations literally erased all evidence of former glaciations on the earth surface, but the full glacial history of the Quaternary is still preserved in the deep-sea sediments.

This glacial "evidence clearing" process may have its parallel in other exposed terrestrial records, including the archaeological record. The evidence of eventual, still unnoticed, discovery and settlement events (especially if they are scarce and/or intermittent) may remain undetected or may have been "cleared" by erosion and/or by further, more permanent and powerful cultures and civilizations, but may still be preserved in sedimentary archives such as lake sediments and peat bogs.

3.1.2 The case of Easter Island

From the above discussion, the usefulness of paleoecology on Easter Island seems clear. First, because paleoecology is able to unravel ecological trends of terrestrial ecosystems, notably vegetation shifts and forest clearing, through time and their respective natural and anthropogenic drivers. Second, because paleoecology can provide evidence on past climates, ecosystems, and human activities unavailable from other fields of research such as archaeology and anthropology. Third, because paleoecological evidence is less sensitive to the phenomenon of evidence clearing, as deep lake/swamp sediments are less accessible to superficial erosion and destruction agents and, hence, the evidence contained in them may be better preserved. This does not mean that paleoecology is superior to other disciplines studying the past but it may furnish useful complementary evidence that should not be neglected. In the specific context of Easter Island, important specific research questions that paleoecological evidence may able to address are the following:

- Is there evidence of human presence before archaeological colonization?
- Did climate remain constant during Easter Island's prehistory?
- If climate had changed, did it experience gradual long-term changes, abrupt shifts, or both?

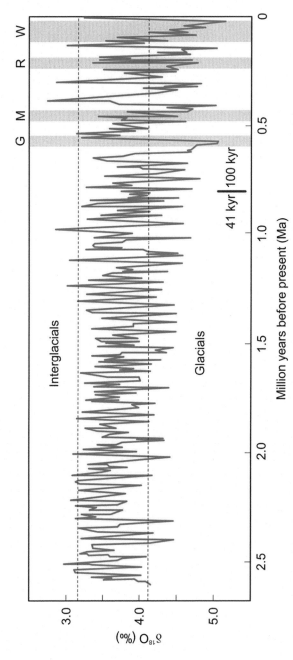

Fig. 3.2 Full Pleistocene glacial-interglacial sequence as recorded in the Atlantic deep-sea core DSDP Leg 607 (Raymo, 1994). *Dotted lines* indicate the threshold values to define glacial (below) and interglacial (above) phases. Vertical gray bands indicate the formerly defined glaciations for Central Europe, based on continental geomorphological evidence (Köppen and Wegener, 1924). G, Günz; M, Mindel; R, Riss; W, Würm.

- Was Easter Island affected by well known, global and quasiglobal climatic shifts of the last millennium such as the Medieval Climatic Anomaly (MCA) and/or the Little Ice Age (LIA)? If so, how? If not, why?
- Which was the most changing climatic parameter: temperature or moisture balance?
- Did eventual climatic shifts affect fire incidence, ecosystem dynamics, deforestation, human activities, and/or fire-vegetation-human interactions?
- Did natural environmental forcings other than climate (e.g., volcanism and tsunamis) affect Easter Island's prehistorical developments?

Most of these questions may be surprising for many paleoecologists who are familiar with worldwide climatic variability during the last millennium (Jones et al., 2001; Bradley et al., 2003) and consider this variability as the norm when analyzing socioecological processes (Goudie, 2006; Roberts, 2014). Pacific islands and archipelagos have also been submitted to this environmental variability that has affected cultural developments (Allen, 2006; Nunn, 2007). Therefore, from a paleoecological perspective, the questions about climate changes and their eventual socioecological consequences might be reframed into: why should Easter Island be different? The truth is that the mainstream archaeological research and, surprisingly, several paleoecological studies have systematically ignored or explicitly denied the occurrence of climate shifts significant enough to affect human affairs during Easter Island's prehistory, as it has occurred elsewhere, including many Pacific islands. However, exceptions exist and several authors advanced the hypothetical possibility of relevant prehistoric climatic shifts able to influence ecological and cultural developments on Easter Island. The following section briefly summarizes these initial speculations, as an introduction to further chapters, where sound evidence for climate change and its socioecological consequences is disclosed in more detail.

3.2 Initial proposals of prehistoric climate change on Easter Island

As soon as the pioneering paleoecological studies of Flenley and coworkers (Flenley and King, 1984; Flenley et al., 1991) confirmed Mulloy's (1974) ideas about the occurrence of an island-wide deforestation (Chapter 4), the first speculations on a potential role for climate change emerged. In these first palynological surveys, Flenley and his colleagues suggested the possibility of several climatic shifts during the last millennia (Chapter 4) but they

did not consider them of sufficient intensity to cause island-wide deforestation. Indeed, Bahn and Flenley (1992) considered that if Easter Island's
forests had survived the pronounced and extended fluctuations in temperature and moisture balance occurred between the Late Pleistocene and the
Middle Holocene, including the Last Glacial Maximum (LGM), it would
be unreasonable to assume that they could have been annihilated by the
comparatively lower climatic variability of the Late Holocene. Slightly later,
McCall (1993) noted that Bahn and Flenley (1992) did not mention that
deforestation roughly coincided with the multicentennial colder and drier
phase known as the Little Ice Age (LIA). McCall (1993) explicitly avoided
a deterministic approach and suggested that the LIA should not be viewed
as a decisive deforestation driver but as a new factor to be considered in the
explanation of social and cultural developments. According to this author,
deforestation could have occurred with or without the LIA influence but
subsequent sociocultural development on a treeless island may have been
affected by climatic conditions different from those existing during the forested phase. McCall (1993) also mentioned the possibility of some influence
of ENSO variability, given the geographical position of Easter Island, but he
did not develop this argument further. Orliac and Orliac (1998) considered
that a severe drought caused by the ENSO activity should not be ruled out
as a potential explanation for forest clearing. Haberle and Chepstow-Lusty
(2000) noted the chronological correspondence of the period of moai
building with a phase of low ENSO activity, and the coincidence of the
environmental stress and the collapse of the Rapanui culture with a period
of intense ENSO activity (Fig. 3.3). However, in contrast to McCall (1993)
and Orliac and Orliac (1998) proposals, deforestation was not associated
with a specific climatic trend.

Hunter-Anderson (1998) reinterpreted the pioneering pollen records of
Flenley and coworkers (Chapter 4) in climatic terms and developed a geo-
climatic model to explain deforestation. According to this author, a climatic
shift occurred prior to human settlement (3000 BP) that increased climatic
variability leading to frequent droughts, hotter summers, and colder winters.
This would have favored faster-growing and faster-reproducing plants such
as grasses, to the detriment of trees. When the Polynesian settlers arrived,
during the Little Climatic Optimum (1200–600 BP), the island was mostly
grass covered and the few remaining trees and shrubs grew on protected
lowland sites. Palms did not go extinct until the LIA owing to cooler climates and severe droughts, which also caused a reduction in agricultural
output and a population decline (Hunter-Anderson, 1998). According to
Nunn (2000) and Nunn and Britton (2001), the transition from the Little

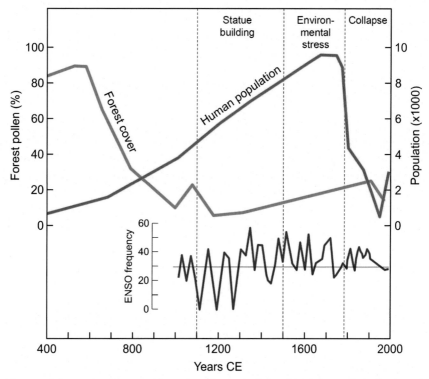

Fig. 3.3 Relationship between the cultural periods defined by Bahn and Flenley (1992) and the ENSO frequency, measured in number of El Niño events per 100 years. *(Redrawn and modified from Haberle, S.G., Chepstow-Lusty, A., 2000. Can climate influence cultural development? A view through time. Environ. Hist. 6, 349–369.)*

Climatic Optimum (LCO) or Medieval Warm Period (MWP) (1200–700 cal yr BP; 750–1250 CE) to the LIA (600–100 cal yr BP; 1350–1800 CE) was characterized, in the Pacific basin, by a drop in sea levels, a temperature decline, and a precipitation increase (due to increasing El Niño frequency) known as the "AD 1300 event," an event whose occurrence has been questioned (Allen, 2006). According to Nunn (2000) and Nunn and Britton (2001), Easter Island was colonized during the MWP, when warm/dry climates and higher sea levels would have favored long-distance voyaging across the Pacific (see also Goodwin et al., 2014). During the AD 1300 event, rapid cooling and uncommonly heavy precipitation could have devastated terrestrial resources, especially the agricultural system that existed during the MWP, thus initiating its depletion. On Easter Island, this would have led to increasing competition for resources, violence and cave-dwelling development during the 15th or 16th centuries, as well as the end of the moai production and the destruction of the existing ahu (Nunn, 2000).

This author suggested that deforestation was likely caused by climate change and mentioned the former hypothesis of Hunter-Anderson (1998) about the possibility of forest retraction being already in progress before Polynesian arrival. Several years later, these authors gathered their respective views and prioritized the AD 1300 event as a regional Pacific phenomenon and the most likely cause for Easter Island's cultural collapse (Nunn et al., 2007). In the context of this environmentally deterministic proposal, the LIA cooler and drier climates could have magnified the socioecological response to the AD 1300 crisis and favored the rapid decline of the endemic palm, the onset of social conflicts, statue toppling, and the shift from the moai cult to the Birdman cult by 1500 CE (Nunn, 2007).

The common problem of all the above proposals about climatic shifts is that, by the time they were suggested, there was no sound and independent evidence of prehistoric climate change on Easter Island and all were speculations and generalizations based on paleoclimatic data from other Pacific areas, often belonging to climate systems very different from those controlling Easter Island's subtropical climate (Fig. 1.5). Examples are the temperate New Zealand (38–46°S and at > 7000 km ESE from Easter Island), in the case of Hunter-Anderson (1998), or an array of tropical and temperate continental localities ranging from North to South America, East Asia, and Australasia, in the case of Nunn (2000) and Nunn and Britton (2001).

The potential mechanisms driving the assumed climate changes, notably the influence of the ENSO activity, were also questioned. For example, as explained in the first chapter (Section 1.2), several analyses of historical climatic series did not support the occurrence of interannual variability in temperature or precipitation that could be associated to the ENSO cyclicity (MacIntyre, 2001; Genz and Hunt, 2003; Caviedes and Waylen, 2011). A different method was attempted by Junk and Claussen (2011), who used precipitation data from the period 1987 to 2005 to feed a climatic model and simulate past climates for the period 800–1750 CE. These authors did not find a direct influence of the ENSO phenomenon, the MWP, or the LIA on their simulations of past climates and concluded that large-scale climatic changes in the oceanic region around Easter Island might be too small to explain strong vegetation changes on the island during the last millennium. However, Stenseth and Voje (2009) found a chronological coincidence between the period of deforestation—between 1250 and 1650 CE, as taken from Hunt (2007)—and the most intense ENSO activity of the last millennium (Fig. 3.4), which was interpreted by these authors in terms of increased deforestation for procuring more wood to build better and larger

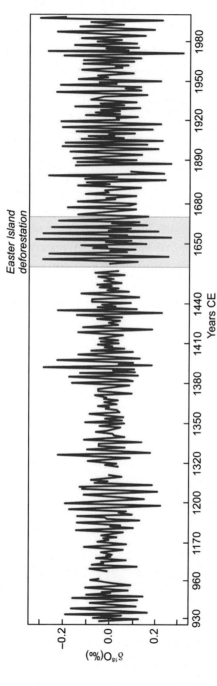

Fig. 3.4 ENSO variability during the last millennium, as deduced from oxygen isotope ratio ($\delta^{18}O$) anomalies in coral records from the Palmyra Atoll (Fig. 1.5). *(Redrawn and modified from Stenseth, N.C., Voje, K.L., 2009. Easter Island: climate change might have contributed to past cultural and societal changes. Clim. Res. 39, 111–114.)*

canoes to gain access to previously inaccessible fishing grounds. Further declines in marine resources may have compelled Easter Islanders to enhance forest clearing to expand arable land, resulting in total island deforestation (Stenseth and Voje, 2009). Again, the main problem was that ENSO reconstructions used by these authors were not based on Easter Island's paleoclimatic archives but on coral records from a tropical Pacific island, the Palmyra Atoll, situated at about 6°N and > 6500 km NW from Easter Island (Cobb et al., 2003) and, therefore, submitted to different climate controls (Fig. 1.5). The only exception is the study developed by Mucciarone and Dunbar (2003) on a coral record from Easter Island already explained in Section 1.2. Unfortunately, this record was limited to the second half of the 20th century and could not inform on prehistoric trends and developments.

This mostly speculative perspective on the potential influence of eventual climate changes on Easter Island's prehistory remained until the first decade of the 21st century. The first local paleoecological evidence of influential climatic shifts and their potential effects on prehistoric sociocultural developments of Easter Island started to be recognized some decades after the first pioneering palynological surveys by Flenley and his collaborators (Chapter 6). The rest of this chapter is dedicated to listing the sites studied and the methods used in the paleoecological study of Easter Island, based on all that has been published until 2019.

3.3 Coring sites

As mentioned in Chapter 1, most paleoecological studies on Easter Island have been performed using peat and lake sediments from the only three permanent waterbodies: Rano Aroi, Rano Kao, and Rano Raraku (Figs. 1.1 and 1.6). These sediments are the archives where organic and inorganic proxies for past vegetation, landscape, and environmental reconstruction have accumulated through time. This section briefly describes the main sedimentary features of the three sites, the cores obtained to date and their chronologies, and the main analyses performed in these cores, as a methodological introduction of the detailed data analysis of the following chapters and their interpretation in terms of past climatic, vegetation and landscape shifts.

3.3.1 Rano Aroi

Rano Aroi is a swamp (mire) of 150 m diameter situated at 430 m elevation near the Terevaka summit (Figs. 1.1 and 1.6). The water level is controlled by groundwater inputs subjected to the influence of seasonal variations in precipitation and human extraction, through the construction of an artificial outlet in the 1960s (Herrera and Custodio, 2008). The aquatic vegetation is dominated by *Scirpus californicus, Polygonum acuminatum,* and ferns of the genera *Asplenium, Vittaria,* and *Cyclosorus,* whereas the surrounding area is covered by grasslands and a small *Eucalyptus* forest planted during the 1960s (Zizka, 1991) (Fig. 1.11). The mire infilling is predominantly peat and is at least 16 m deep in the center, which may represent an extrapolated age of approximately 70,000 yr BP (Margalef et al., 2013, 2014). However, the deposit could be deeper and older as none of the existing sedimentary records reached the bedrock. Flenley et al. (1991) described this organic accumulation as a mixture of coarse detritus and finer material intermingled with layers of spongy monocotyledonous peat and brown clay. Peteet et al. (2003) reported the occurrence of several types of peats and organic clays with fibrous material. A similar composition was described by Horrocks et al. (2015). Margalef et al. (2013, 2014) did a detailed lithological study of the stratigraphic sequence and distinguished four main organic facies: (A) reddish peat of sedges and *Polygonum;* (B) granulated muddy peat of coarse organic fragments, mainly roots, with low terrigenous content; (C) organic mud or dark-brown to black peat, and (D) dark-brown fine-grained peat. Facies B is the better represented throughout the sequence.

3.3.2 Rano Kao

Rano Kao contains the largest lake of the island, with 1250 m diameter, situated at 110 m elevation. This lake is very peculiar as its surface is a mosaic of water and aquatic vegetation in the form of floating mats up to 3–4 m deep overlying the water column, which is about 10 m deep near the center (Figs. 1.1 and 1.6). This configuration determines the existence of two different paleoecological archives: the superficial peaty floating mat and the more clastic bottom lake sediments accumulated below the water column. The floating mats are dominated by the characteristic semiaquatic species of the island, *Scirpus californicus* and *Polygonum acuminatum,* together with another sedge, *Pycreus polystachyos* (Zizka, 1991). The oldest ages recorded to date in the floating mat correspond to the last millennium (Gossen, 2007; Horrocks et al., 2013). It has been suggested that part of the floating mat would have been occasionally submerged and mixed with the uppermost bottom sediments thus causing chronological anomalies, typically age inversions, in the sedimentary sequence (Butler et al., 2004). A significant amount of archaeological sites has been found within and around Rano Kao, including the ancient village of Orongo, formed by stone houses, which is one of the most important and well-preserved archaeological complexes of the island (Robinson and Stevenson, 2017). The maximum depth of lake sediments recorded thus far is about 21 m (but they should be deeper) and the maximum age is 34,000 cal yr BP, although the more frequent ages for the base of the cores are around 20,000 cal yr BP (Gossen, 2007, 2011; Horrocks et al., 2013). The Kao lake sediments have been described as coarse organic detritus derived from aquatic and catchment vegetation, with a basal layer of coarse detritus and clay (Flenley and King, 1984; Flenley et al., 1991; Horrocks et al., 2013).

3.3.3 Rano Raraku

Rano Raraku contains a small and shallow lake of 300 m diameter and about 2 m depth, situated at 80 m elevation (Figs. 1.1 and 1.6). Hydrologically, the lake is closed, with no surface outlet, and is used by humans as a freshwater source for consumption and irrigation. The main water inputs are rainfall and catchment runoff (Herrera and Custodio, 2008). The water level is in-fluenced by periodic water extraction. The maximum depth recorded in modern times is about 3 m (Sáez et al., 2009), but in some years, the lake may dry completely. The aquatic vegetation is dominated by *Scirpus californicus*, which forms a more or less continuous floating belt (in part rooted in lake sediments) in the eastern margin of the lake (Fig. 1.6), where the input of terrigenous materials from the catchment is higher and more continuous. Rano Raraku is one of the most emblematic sites of the island, as it was the quarry where the moai were carved. Many of these stone statues, some of them unfinished, still remain in the east side of the crater where most moais were carved (Fig. 2.10). The sedimentary infilling is at least 14 m deep in the center of the lake, which corresponds to an age of 34,000 cal yr BP (Sáez et al., 2009). These sediments have been described as a mixture of coarse and fine detritus originated from lake and catchment vegetation, intermingled with layers of gyttja, clay and mud, and, eventually, volcanic ashes (Flenley et al., 1991; Horrocks et al., 2012a). A more detailed sedimentological study reported that the Raraku sediments were dominated by organic matter (70%–90% of the total weight), with variable amounts of terrigenous mineral particles from the catchment rocks and pyritic precipitates. The organic matter is a mixture of plant remains from the catchment and, in a lower proportion, autochthonous organic matter derived from lake production (Sáez et al., 2009). Four main sedimentary facies have been distinguished (from the base to the top): (1) laminated, dark gray-reddish organic-rich silts and mud; (2) laminated and massive brownish organic mud; (3) brown-reddish massive or banded peaty sediment, composed mainly of sedge remains; and (4) peat and silty clay (Fig. 3.11).

3.4 Cores retrieved and main proxies studied

Cores from the Aroi, Kao, and Raraku sediments have been retrieved at different parts of these ranos, using a variety of coring equipment and extrusion techniques (Fig. 3.5 and Table 3.1), see Glew et al. (2001) for details on these coring methods. Fig. 3.6 shows the location of all cores obtained to date with published results, whereas Table 3.1 displays the main features of each core and the main paleoclimatic and paleoecological proxies studied. The results obtained in these studies and their interpretation in the framework of the climatic and socioecological developments of Easter Island's prehistory are disclosed in detail in Chapters 4–6. Details on these analytical methods and the most usual statistical and plotting techniques for data management can be found in Last and Smol (2001a,b), Smol et al. (2001a,b), and Birks et al. (2012).

The progress of paleoecological research on Easter Island has not been continuous but has occurred in several phases. In this book, three main stages have been considered that are the basis for the following chapters.

Fig. 3.5 Coring devices utilized in Easter Island's ranos in the 2006 and 2008 campaigns (Table 3.1). (A) UWITEC platform at Rano Raraku (2006). (B) UWITEC platform at Rano Aroi (2006). (C) Russian corer at Rano Kao (2008). *(Photos by A. Sáez (A), A. Moreno (B), and O. Margalef (C).)*

Table 3.1 Characteristics of radiocarbon-dated peat and lake sediment cores obtained in Easter Island, according to the original references.

Rano Aroi (27°05′37.37″N—109°22′26.50″; 433 m elevation)

Core	Water depth (m)	Core length (m)	Year	Coring device	Main proxies studied	References	Publication lag (years)
ARO1	0.00	~11.50	1977	Russian	Lithostratigraphy, elemental analysis, pollen and spores, charcoal	Flenley (1979b), Flenley and King (1984), and Flenley et al. (1991)	7
ARO 06–01	0.00	13.90	2006	UWITEC	Facies description, mineralogy, elemental analysis, stable isotopes, plant and animal macrofossils, pollen and spores	Margalef (2014) and Margalef et al. (2013, 2014)	7
ARO 08–02	0.00	4.00	2008	Russian	Facies description, mineralogy, elemental analysis, stable isotopes, plant and animal macrofossils, pollen and spores	Margalef (2014), Margalef et al. (2013, 2014), and Rull et al. (2015)	7
RA2	0.00	8.00	1997	Livingstone	Lithostratigraphy, diffuse spectral reflectance, pollen and spores	Peteet et al. (2003)	16
RA	0.00	4.00	2009	Russian	Lithostratigraphy, magnetic susceptibility, plant and animal macrofossils, charcoal, pollen and spores, biosilicates, phytoliths, starch	Horrocks et al. (2015)	6
RAI[a]	0.00	2.11	2009	Russian	Lithostratigraphy, magnetic susceptibility, plant and animal macrofossils, charcoal, pollen and spores, biosilicates, starch	Horrocks et al. (2015)	6

Rano Kao (27°11′12.57″N—109°26′06.75″; 107 m elevation)							
Core	Water depth (m)	Core length (m)	Year	Coring device	Main proxies studied	References	Publication lag (years)
KAO1	0.00	~11.00	1977	Russian	Lithostratigraphy, elemental analysis, pollen and spores, charcoal	Flenley (1979b), Flenley and King (1984), and Flenley et al. (1991)	7
1	ND	20.00	2009	Russian + Livingstone	Lithostratigraphy, magnetic susceptibility, plant and animal macrofossils, charcoal, pollen and spores, starch	Horrocks et al. (2013)	4
2	ND	~6.00	2009	Russian + Livingstone	Lithostratigraphy, phytoliths, pollen and spores, starch	Horrocks et al. (2012b)	3
3	ND	~5.00	2009	Russian + Livingstone	Lithostratigraphy, phytoliths, pollen and spores, starch	Horrocks et al. (2012b)	3
4	ND	~12.00	2009	Russian + Livingstone	Lithostratigraphy, phytoliths, pollen and spores, starch	Horrocks et al. (2012b)	3
5	ND	~7.00	2009	Russian + Livingstone	Lithostratigraphy, phytoliths, pollen and spores, starch	Horrocks et al. (2012b)	3
KAO2	10.50	20.85	1983	Russian	Lithostratigraphy, elemental analysis, pollen and spores, charcoal	Butler and Flenley (2001, 2010), Butler et al. (2004), and Flenley (1996)	18
KAO3 (KAO05-3A)	10.50	21.50	2005	Russian + Livingstone	Magnetic susceptibility, oxygen isotopes, pollen and spores	Gossen (2007, 2011)	2
KAO08-03	ND	2.20	2008	Russian	Lithostratigraphy, pollen and spores, charcoal, nonpollen palynomorphs (NPP)	Rull et al. (2018) and Seco et al. (2019)	10

Continued

Rano Raraku (27°07'19.79"—109° 17'20.66"; 80 m elevation)

Core	Water depth (m)	Core length (m)	Year	Coring device	Main proxies studied	References	Publication lag (years)
RRA3	0.00	~12.00	1977	Russian	Lithostratigraphy, elemental analysis, pollen and spores, charcoal	Flenley (1979b), Flenley and King (1984), and Flenley et al. (1991)	7
RRA4	2.80	~17.20	1983	Russian	Lithostratigraphy, mineralogy, elemental analysis	Flenley et al. (1991)	8
RRA5	ND	13.40	2005	Livingstone	Pollen and spores	Azizi and Flenley (2008)	3
SW	2.00	~3.40	1990	Piston	Magnetic properties, plant and animal microfossils (pollen, cladocera, ostracoda, diatoms), pigments	Dumont et al. (1998)	8
1	6.00	~1.00	1998	Gravity	Lithostratigraphy, magnetic susceptibility, organic matter, charcoal, pollen and spores	Mann et al. (2003, 2008)	10
2	6.00	~1.00	1998	Gravity	Organic matter	Mann et al. (2003, 2008)	10
RAR01	ND	~5.70	2006	UWITEC	Magnetic susceptibility, facies description, elemental analysis, mineralogy	Sáez et al. (2009)	3
RAR02	ND	~6.80	2006	UWITEC	Magnetic susceptibility, facies description, elemental analysis, mineralogy	Sáez et al. (2009)	3

RAR03	ND	~13.80	2006	UWITEC	Magnetic susceptibility, facies description, elemental analysis, mineralogy, plant and animal macrofossils, pollen and spores	Sáez et al. (2009), Cañellas-Boltà (2014), and Cañellas-Boltà et al. (2012, 2014, 2016)	3
RAR04	ND	~6.00	2006	UWITEC	Magnetic susceptibility, facies description, elemental analysis, mineralogy	Sáez et al. (2009)	3
RAR05	ND	~13.20	2006	UWITEC	Magnetic susceptibility, facies description, elemental analysis, mineralogy	Sáez et al. (2009)	3
RAR07	ND	~12.80	2006	UWITEC	Magnetic susceptibility, facies description, elemental analysis, mineralogy, plant and animal macrofossils, pollen and spores	Sáez et al. (2009), Cañellas-Boltà (2014), and Cañellas-Boltà et al. (2012, 2014, 2016)	3
RAR08	ND	~12.60	2006	UWITEC	Magnetic susceptibility, facies description, elemental analysis, mineralogy	Sáez et al. (2009), Cañellas-Boltà (2014), and Cañellas-Boltà et al. (2013)	3
Lake	2.60	2.25	2009	Livingstone	Lithostratigraphy, plant macrofossils, pollen and spores, phytoliths, starch	Horrocks et al. (2012a)	3

ND, no data.
[a] Rano Aroi Iti (see Figs. 1.9 and 3.6).

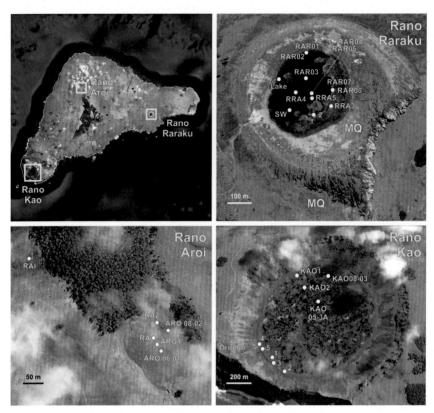

Fig. 3.6 Google-Earth images of the paleoecological sites of Easter Island indicating all cores retrieved to date with published results, which constitute the basis for the EIRA radiocarbon database discussed in the text. MQ, Moai Quarry (crater slopes where moai were carved). *(Modified from Rull, V., 2016a. The EIRA database: last glacial and holocene radiocarbon ages from Easter Island's sedimentary records. Front. Ecol. Evol. 4, 44.)*

These phases are (Fig. 3.7): the pioneering phase (Chapter 4), the transitional phase (Chapter 5), and the revival phase (Chapter 6). This chronological subdivision has been based on the combination of fieldwork activity and the publication of the corresponding results, which is typically lagged by several years due to the complexity and time-consuming nature of paleoecological work. On Easter Island, the time lag between lake/swamp coring and the publication of the corresponding paleoecological results is of 6 years, on average, ranging from 3 to 18 years (Table 3.1). The pioneering phase began in 1977, with the first systematic coring of Rano Aroi, Rano Kao, and Rano Raraku by John Flenley and his team, and ended in the early 1990s, with the seminal publications by Flenley et al. (1991) and Bahn

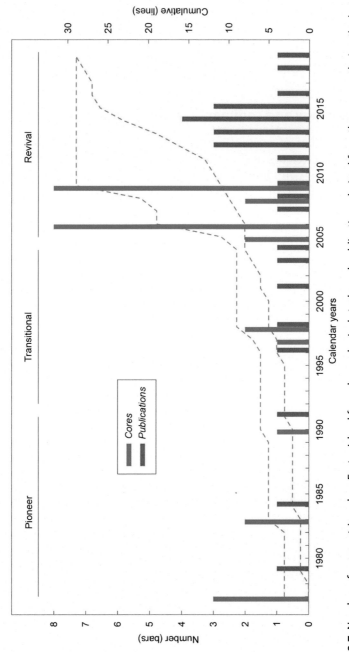

Fig. 3.7 Number of cores retrieved on Easter Island for paleoecological study and publications derived from these cores during the last four decades (1977–2019). Only publications with original data (Table 3.1) on the cores are considered; further reviews and commentaries are excluded. Bars represent the actual number of cores and papers (left scale), and dotted lines are the cumulative trends (right scale).

and Flenley (1992). The intermediate phase has been considered here a phase of paleoecological impasse, due to the comparatively low coring and publishing activity between about 1993 and 2004. The revival phase started with a considerable increase in lake/swamp coring between 2005 and 2009, which led to a significant publication increase several years later due to the abovementioned coring-publication lag.

3.5 Chronology and sedimentary patterns

A variable number of samples from these cores have been submitted to standard or accelerator mass spectrometry (AMS) radiocarbon dating (Björk and Wohlfarth, 2001) to develop age-depth models using several methods, from simple linear interpolation to Bayesian inference (Blaauw et al., 2018). All the raw dates and, whenever possible, their calibration to calendar years using the online software CALIB 7.1 (www.calib.org) and the Southern Hemisphere calibration database SHCal13 (Hogg et al., 2013), have been gathered in a database called Easter Island Radiocarbon Ages (EIRA), with 288 entries (Rull, 2016a), which is periodically updated and is publicly available at the NOAA National Climatic Data Center (www. ncdc.noaa.gov/paleo-search/study/19805). The total number of calibrated/ recalibrated dates is 270 (excluding unreliable ages of > 50 cal kyr BP), of which 208 (77%) correspond to the Holocene (11,700–0 cal yr BP) and the remaining 62 (23%) are Late Pleistocene (48,000–17,500 cal yr BP). The site with more dates is Rano Kao (120 dates; 44%), followed by Rano Raraku (85 dates; 32%) and Rano Aroi (65 dates; 24%). Overall, the time interval with more dates is the last millennium, with a more or less maintained exponential decrease until the Pleistocene, except for a small increase in the Early Holocene (Fig. 3.8). Late Pleistocene dates are scarce, except for the Lateglacial interval (18–12 cal kyr BP). The interval with fewer dates is 18–22 cal kyr BP, which corresponds to the LGM (Azizi and Flenley, 2008). Individual sites follow similar trends with some variations. In Rano Aroi, Late Pleistocene dates are more frequent, with noticeable gaps in the LGM and the Middle Holocene. Late Pleistocene ages are almost inexistent in Rano Kao, where most dates are concentrated in the Late Holocene, with very few in the Middle Holocene. In Rano Raraku, the Middle Holocene is better represented than in other sites and the more abundant Pleistocene ages correspond to the Lateglacial.

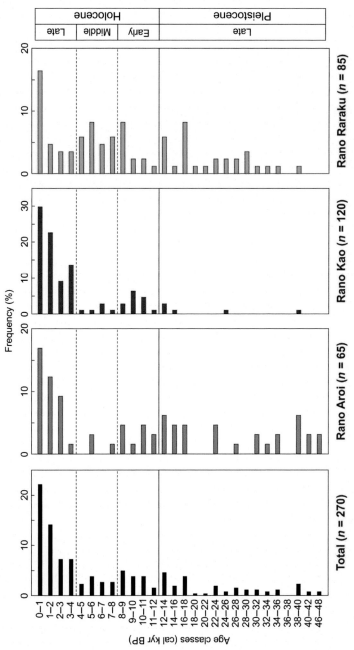

Fig. 3.8 Age-class frequency histograms using calibrated median probably values of the EIRA database (Rull, 2016a). See Appendix 3.1 for raw data. Unreliable ages of > 50 ¹⁴C kyr BP have not been used. Geochronological units according to Litt and Gibbard (2008) and Walker et al. (2019).

Some interesting trends may be observed when all dates obtained in each sedimentary body are put together in simple age-depth biplots (Fig. 3.9). These trends will be tentatively discussed here in a general manner and will be analyzed in more detail, case by case, in further chapters. The first general observation is that the three ranos have disparate sedimentary trends. In Aroi, the sediments surveyed started to accumulate by 50 cal kyr BP in a more or less continuous manner until about 30 cal kyr BP, when a sedimentary gap elapsing about 15 cal yr BP occurred. The sedimentation reinitiated by 15 cal kyr BP and has been continuous until today, although accumulation rates seem to have been slower than before. Interestingly, the uppermost meter shows ages between modern and > 20 cal kyr BP, which may be consistent with vertical mixing of sediments caused by human disturbance in the search for freshwater. For example, it is known that, in the 1920s, teams of bullocks were used to drag the sediments and build a dam in an attempt to increase the water retention capacity of the Aroi swamp (Flenley et al., 1991).

The Kao sedimentary patterns are more complex because of the already mentioned occurrence of the floating peaty mat embracing roughly the last millennium. This creates an intricate sedimentary gradient between the center and the margins of the lake. In the center, three layers (floating mat, open water, and lake sediments) are clearly defined, whereas the margins are characterized by massive peats of similar nature of the floating mat. In between, there is a complex mixture of layers of peats, water, and sediments of different types and ages. Perhaps the most intriguing phenomena here is the occurrence of a negative exponential trend from about 15 to 3 cal kyr BP in the bottom sediments (22–16 m) and the further stabilization in values up to 3–4 cal kyr BP from 16 m to the surface (Fig. 3.9). This would imply huge average accumulation rates of 0.5 cm/yr or more during the last millennia, which is very different from the situation in other ranos. To explain this situation, as well as the formation of the floating peat mats, Horrocks et al. (2013) suggested the following sequence of events (Fig. 3.10): (i) a phase of dry climates and low lake levels (prior to 3.5 cal kyr BP), during which sedimentation ceased; (ii) a moisture increase by 3.5 cal kyr BP leading to the formation of a thick (c.16 m) peat layer on older (pre-Holocene) lake sediments; and (iii) a further lake-level increase that would have fragmented this peat layer into two parts, the basal one remaining attached to the older lake sediments and the upper layer migrating upward with the rising water level to form the present network of floating mats.

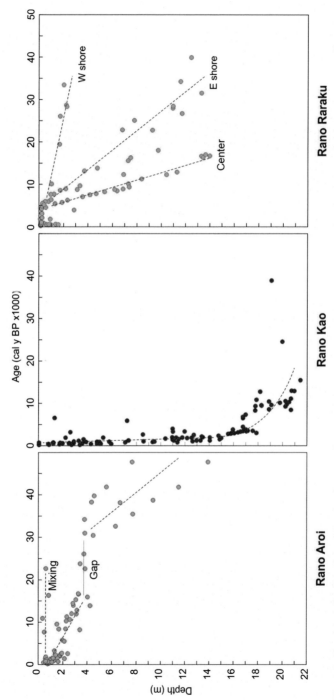

Fig. 3.9 Graphical display of the EIRA database (Rull, 2016a), using the calibrated median probability values. Unreliable ages of > 50 ^{14}C kyr BP have not been used. See Appendix 3.1 for raw data.

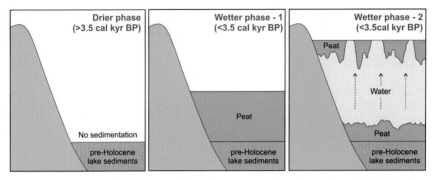

Fig. 3.10 Schematic diagram to explain the age constancy (up to 3–4 cal kyr BP) of radiocarbon ages from the uppermost 16 m of Raraku sediments and the formation of present floating mats. *(Redrawn from Horrocks, M., Marra, M., Baisden, W.T., Flenley, J., Feek, D., González-Nualart, L., Haoa-Cardinali, S., Edmunds Gorman, T., 2013. Pollen, phytoliths, arthropods and high-resolution 14C sampling from Rano Kau, Easter Island: evidence for late quaternary environments, ant (formicidae) distributions and human activity. J. Paleolimnol. 50, 417–432.)*

The Raraku ages show a greater dispersion but there is a structure in this dispersion that is consistent with a spatial pattern (Fig. 3.9). Indeed, accumulation rates are higher in the center, intermediate in the eastern margin, and lower in the western margin. Higher sediment accumulation in the lake depocenter (Fig. 3.11) is not surprising and is most probably enhanced by the focusing effect, i.e., the reworking of sediments from shallower zones and their further sedimentation in the deeper zones of the lake (Likens and Davis, 1975). The difference between eastern and western margins may be explained by higher sedimentary releases from the eastern crater wall, as manifested today in the asymmetric growing of the *Scirpus* floating mats, which grow toward the center of the lake following the predominant sedimentation direction (Fig. 3.5). It should also be noted that the eastern wall was an important part of the moai quarry and, therefore, sediment delivery to the lake would have been magnified by human activities during the prehistory. The internal architecture of the Raraku sediments affects the age-depth models, as the same depth could have different ages in different places of the lake due to the configuration of the sedimentary units (Fig. 3.11). This is expected to be true also for Rano Aroi and Rano Kao but, unfortunately, no similar studies of the internal structure of the whole sedimentary body exist for these sediments. A feature that is not evident in the general Raraku age-depth model (Fig. 3.9), due to the lack of resolution, is the

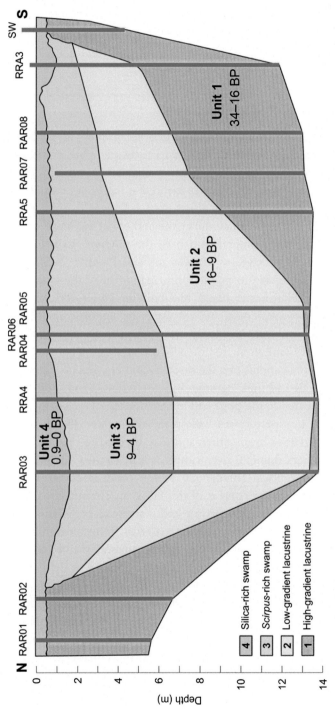

Fig. 3.11 N-S transect showing the sedimentary units and the internal architecture of Lake Raraku sediments. The cores used to reconstruct the units are shown in *blue lines* (see Fig. 3.6 for location). Unit ages are given in cal kyr (yr in the case of Unit 4) BP. *(Modified from Sáez, A., Valero-Garcés, B., Giralt, S., Moreno, A., Bao, R., Pueyo, J.J., Hernández, A., Casas, D., 2009. Glacial to Holocene climate changes in the SE Pacific. The Raraku Lake sedimentary record (Easter Island, 27°S). Quat. Sci. Rev. 28, 2743–2759.)*

occurrence of an extended Holocene sedimentary gap of several millennia (Fig. 3.11), which is also a potential handicap for paleoecological studies. This gap is well documented in several cores, as is explained in more detail in Chapter 6 (Section 6.1.1).

3.6 Other paleoecological archives

Besides lake and swamp sediments, other paleoecological archives have been investigated on Easter Island, although less intensively. One is the already mentioned coral record from Ovahe beach (Section 1.2), which is the only study of this type published from Easter Island that has been produced using paleoecological methods. Although this record is limited to modern times (i.e., the second half of the 20th century), the authors of the study suggested that the coral records of Easter Island have the potential to span at least 250 years, and possibly 400 years (Mucciarone and Dunbar, 2003). If true, these records might be able to provide information since the latter prehistoric times onward. On Easter Island, reef-building corals occur mostly in the northern and western coasts and are scarce or absent in the SE coasts (Fig. 3.12), due to lower temperatures and stronger wave action (Hubbard and García, 2003). Owing to the remoteness, the subtropical climate, and the general oceanographic conditions, the diversity and physical development of Easter Island's corals and their communities are reduced and look like depauperate, as compared to the huge and luxuriant tropical reefs (Glynn et al., 2003). The raised marine terraces widespread throughout the tropics date back to c.130 kyr BP, or even earlier, and may be used to reconstruct SST and other parameters (rainfall, river runoff, ocean circulation, and ENSO activity) since last interglacial (Bradley, 2015). The corals around Easter Island are more modest but, in the context of this book, an eventual record of the last millennium would be very welcome and is worth to be pursued.

The study of soil profiles and selected materials from some archaeological sites have also provided paleoecological information, mainly on the nature of ancient forests and deforestation timing. In several of these archives, Mann et al. (2003) and Mieth and Bork (2005, 2010, 2015, 2017) found evidence of slash and burn activities, agroforestry, forest clearing and regeneration, soil erosion, and other ecological processes that will be described in more detail in the corresponding chapters. In other dryland soils and archaeological sites, Horrocks and Wozniak (2008) and Horrocks et al. (2016) found evidence for forest clearing and cultivation of Polynesian crops that will also be further disclosed. The already mentioned archaeological sites

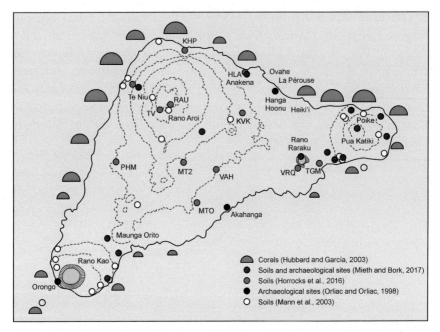

Fig. 3.12 Distribution of corals (*green* semicircles) around Easter Island. The size of semicircles is a rough indication of the abundance and cover of coral colonies (Hubbard and García, 2003). *Red*, *blue*, *white*, and *black* dots mark the sites where detailed edaphic studies were carried out and some archaeological sites from where paleoecological evidence has been collected (Orliac, 2000; Orliac and Orliac, 1998; Mann et al., 2003; Horrocks et al., 2016; Mieth and Bork, 2003, 2005, 2010, 2015, 2017).

where Orliac (2000), and Orliac and Orliac (1998) identified woody plant taxa from charcoal fragments (Table 5.1) are also worth to be mentioned (Fig. 3.12). The usual proxies utilized in these studies have been soil stratigraphy, with emphasis on carbon layers, charred wood/seeds, root casts, pollen, phytoliths, and starch remains.

Appendix 3.1

Radiocarbon ages from the cores obtained to date in Rano Aroi, Rano Kao, and Rano Raraku. Depth = average sample depth in meters; LCR = lower calibrated range (2σ); UCR = upper calibrated range (2σ); RAPD = relative area under probability distribution; Median = median probability; ND = no data. Radiocarbon laboratory codes: AA, University of Arizona AMS Facility (USA); Beta, Beta Analytic Inc. (USA); CAMS, Center for

Acceleration Mass Spectrometry (USA); NUTA, Department of Geography, Nagoya University (Japan); NZA, Rafter Radiocarbon Laboratory (New Zealand); Poz, Poznan Radiocarbon Laboratory (Poland); SRR, Scottish Universities Research and Reactor Centre (Scotland); ULA, Université Laval (Canada). See Fig. 3.6 for location and Table 3.1 for references and core details. All radiocarbon dates have been calibrated or recalibrated by Rull, (2016a) using the SHCal13 calibration database for the Southern Hemisphere (Hogg et al., 2013).

Rano Aroi

Core ARO1 (Flenley, 1979b; Flenley and King, 1984; Flenley et al., 1991).

Lab code	Depth	Material	^{14}C yr BP	Error	LCR	UCR	RAPD	Median
SRR–2417	0.600	ND	18,870	140	22,407	23,027	1.000	22,697
SRR–2037	1.045	ND	Modern	ND	ND	ND	ND	ND
SRR–1790	1.750	ND	2100	50	1895	2154	0.996	2032
SRR–2418	2.200	ND	7680	80	8309	8597	0.992	8444
SRR–2038	3.145	ND	10,960	130	12,656	13,071	1.000	12,828
SRR–2419	3.800	ND	18,780	250	21,991	23,223	1.000	22,628
SRR–1791	4.450	ND	26,280	300	29,715	30,999	1.000	30,467
SRR–2420	6.250	ND	28,670	950	30,987	34,394	1.000	32,630
SRR–1702	7.650	ND	31,610	720	34,043	37,183	1.000	35,513
SRR–2421	9.300	ND	34,440	1040	36,307	41,067	1.000	38,794
SRR–1353	11.400	ND	37,680	1120	39,783	43,631	1.000	41,875

Core ARO 06-01 (Margalef, 2014; Margalef et al., 2013, 2014).

Lab code	Depth	Material	^{14}C yr BP	Error	LCR	UCR	RAPD	Median
ND	2.380	Pollen extract	2580	30	2490	2643	0.602	2620
ND	2.550	Pollen extract	9460	50	10,501	10,785	0.989	10,648
Poz–31994	2.850	Pollen extract	12,150	60	13,773	14,130	1.000	13,964
ND	3.050	Pollen extract	12,880	70	15,099	15,605	1.000	15,313
ND	3.270	Pollen extract	13,800	60	16,356	16,906	1.000	16,624
ND	3.380	Pollen extract	7440	50	8151	8350	0.895	8238
ND	3.780	Pollen extract	26,960	150	30,789	31,220	1.000	31,013
Poz–31995	4.210	Pollen extract	12,070	60	13,742	14,061	1.000	13,883
Poz–31996	4.315	Pollen extract	34,000	500	36,836	39,554	1.000	38,348
ND	4.530	Pollen extract	35,300	600	38,584	41,119	1.000	39,820
ND	5.520	Pollen extract	37,600	600	40,940	42,769	1.000	41,915
ND	6.620	Pollen extract	33,900	500	36,719	39,416	1.000	38,202
ND	7.600	Pollen extract	45,000	2000	44,973	50,000	1.000	47,797
ND	8.720	Pollen extract	49,000[a]	3000[a]	ND	ND	ND	ND
ND	9.790	Pollen extract	52,000[a]	4000[a]	ND	ND	ND	ND
ND	10.830	Pollen extract	>50,000[a]	ND	ND	ND	ND	ND
ND	11.810	Pollen extract	53,000[a]	4000[a]	ND	ND	ND	ND
Poz–35772	12.880	Pollen extract	>49,000[a]	ND	ND	ND	ND	ND
ND	13.800	Pollen extract	49,000	3000	44,970	49,997	1.000	47,794

[a]Invalid ages for calibration curve SHcal13.

Core ARO 08-02 (Margalef, 2014; Margalef et al., 2013, 2014; Rull et al., 2015).

Lab code	Depth	Material	^{14}C yr BP	Error	LCR	UCR	RAPD	Median
Poz–45907	0.760	Pollen extract	190	30	134	285	0.704	178
Poz–31998	0.850	Pollen extract	270	30	272	325	0.560	287
Beta–397173	1.100	Pollen extract	760	30	635	690	0.766	665
Poz–31999	1.250	Pollen extract	1050	30	898	963	0.680	921
Beta–397174	1.540	Pollen extract	1730	30	1533	1629	0.715	1594
Poz–32000	1.750	Pollen extract	2265	30	2155	2329	1.000	2232
Beta–197175	1.840	Pollen extract	2550	30	2463	2742	1.000	2589
Poz–35774	1.940	Pollen extract	5070	40	5659	5896	1.000	5787
Poz–32002	2.250	Pollen extract	9180	50	10,206	10,432	0.971	10,303
Poz–32003	2.810	Pollen extract	12,420	60	14,115	14,814	1.000	14,432
Poz–32004	3.250	Pollen extract	13,880	70	16,456	17,023	1.000	16,748
Poz–32005	3.750	Pollen extract	30,300	300	33,794	34,786	1.000	34,282

Core RA2 (Peteet et al., 2003).

Lab code	Depth	Material	^{14}C yr BP	Error	LCR	UCR	RAPD	Median
AA–29091	0.350	Plant macroremains	9650	170	10,420	11,358	0.998	10,930
AA–43217	0.425	Plant macroremains	492	43	449	548	0.962	505
AA–43218	0.485	Plant macroremains	6880	230	7289	8068	0.977	7702
AA–43219	0.575	Plant macroremains	1565	33	1345	1515	0.971	1404
AA–29090	0.610	Plant macroremains	1944	42	1731	1931	0.984	1848
AA–43220	0.675	Plant macroremains	1300	100	958	1320	0.998	1163
AA–29089	0.830	Plant macroremains	13,605	90	16,050	16,664	1.000	16,343

Core RA (Horrocks et al., 2015).

Lab code	Depth	Material	^{14}C yr BP	Error	LCR	UCR	RAPD	Median
NZA-37403	0.500	Seeds	−1283	30	312	474	0.980	393
NZA-37402	0.600	Seeds	368	35	486	524	0.995	504
NZA-37425	0.700	Seeds	482	20	662	764	1.000	706
NZA-34429	0.800	Seeds	824	40	ND	ND	ND	ND
NZA-38814	0.900	Seeds	−800[a]	15[a]	ND	ND	ND	ND
NZA-37386	1.000	Plant macroremains	1519	40	1298	1432	0.987	1359
NZA-38872	1.100	Seeds	1369	25	1185	1252	0.515	1251
NZA-39387	1.200	Seeds	1724	40	1512	1707	1.000	1591
NZA-37401	1.300	Seeds	3127	35	3205	3382	0.970	3291
NZA-38873	1.400	Seeds	2314	25	2298	2351	0.583	2310
NZA-34859	1.500	Seeds	8634	45	9490	9671	1.000	9551
NZA-37503	1.655	Bulk sediment	7634	25	8348	8430	1.000	8396
NZA-38878	1.900	Seeds	2599	20	2695	2752	0.642	2718
NZA-34430	2.100	Seeds	4855	45	5463	5647	0.958	5527
NZA-37387	2.300	Seeds/plant macroremains	9927	55	11,196	11,411	0.916	11,295
NZA-37389	2.700	Seeds	10,323	55	11,802	12,180	0.769	12,031
NZA-37388	3.100	Seeds	10,283	50	11,749	12,104	0.963	11,939
NZA-34862	3.400	Seeds	19,814	95	23,520	24,078	1.000	23,803
NZA-37407	3.700	Seeds	21,950	120	25,882	26,433	1.000	26,118
NZA-34726	4.000	Seeds	13,306	75	15,705	16,196	1.000	15,946

[a]Invalid age for calibration curve SHcal13.

Core RAI (Horrocks et al., 2015).

Lab code	Depth	Material	^{14}C yr BP	Error	LCR	UCR	RAPD	Median
NZA-37398	0.600	Seeds	935	35	733	845	0.717	796
NZA-37379	1.100	Seeds	626	40	583	650	0.574	598
NZA-37380	1.600	Seeds	840	40	665	773	1.000	713
NZA-32804	2.110	Seeds	1588	55	1317	1540	1.000	1439

Rano Kao

Core KAO1 (Flenley, 1979b; Flenley and King, 1984; Flenley et al., 1991).

Lab code	Depth	Material	^{14}C yr BP	Error	LCR	UCR	RAPD	Median
SRR-2422	1.900	ND	550	60	456	573	0.841	532
SRR-2423	3.900	ND	1000	70	726	978	1.000	857
SRR-2039	4.850	ND	1040	60	767	987	0.964	888
SRR-2424	7.375	ND	1360	50	1159	1309	0.913	1231
SRR-1351	9.350	ND	990	70	723	967	1.000	851

Core 1 (Horrocks et al., 2013).

Lab code	Depth	Material	^{14}C yr BP	Error	LCR	UCR	RAPD	Median
NZA-38317	1.200	Seeds	890	10	728	772	1.000	750
NZA-34921	1.700	Seeds	309	15	293	327	0.720	315
NZA-38318	2.500	Plant macroremains	674	45	549	663	1.000	606
NZA-40267	2.600	Animal macroremains	3024	193	2745	3608	1.000	3150
NZA-37012	2.700	Seeds/plant macroremains	220	25	142	226	0.756	195

Continued

Core 1 (Horrocks et al., 2013)—cont'd

Lab code	Depth	Material	¹⁴C yr BP	Error	LCR	UCR	RAPD	Median
NZA-38320	2.900	Seeds/plant macroremains	727	70	547	729	1.000	635
NZA-38319	3.300	Plant macroremains	1280	210	731	1543	1.000	1142
NZA-33512	3.700	Seeds	1620	150	1257	1829	0.975	1491
NZA-38324	4.800	Seeds	1029	15	901	932	0.535	906
NZA-37011	7.100	Seeds	918	25	728	806	0.914	773
NZA-34920	8.600	Plant macroremains	1147	25	957	1061	1.000	1010
NZA-38325	9.500	Seeds/plant macroremains	1106	15	928	979	0.984	953
NZA-37003	10.400	Seeds	1788	25	1588	1715	1.000	1652
NZA-40268	10.975	Animal macroremains	2028	149	1608	2316	1.000	1951
NZA-40269	11.000	Animal macroremains	3657	325	3158	4841	1.000	3963
NZA-38879	11.000	Seeds	1578	20	1363	1488	0.967	1412
NZA-40270	11.025	Animal macroremains	3131	319	2465	4090	1.000	3284
NZA-38326	11.100	Seeds	1361	15	1185	1221	0.528	1220
NZA-34914	12.200	Seeds/plant macroremains	1886	30	1708	1838	0.976	1781
NZA-37004	13.500	Plant macroremains	2071	25	1922	2055	1.000	1975
NZA-33261	14.800	Seeds	2361	40	2298	2472	0.838	2341
NZA-37260	15.600	Seeds	2697	35	2738	2849	1.000	2772
NZA-37261	15.800	Seeds	2785	35	2759	2928	0.997	2834
NZA-37262	16.100	Seeds	2869	35	2843	3064	0.990	2931
NZA-37013	16.400	Seeds	3017	25	3020	3236	0.990	3143
NZA-33680	16.600	Seeds	3150	30	3215	3396	1.000	3313
NZA-37007	16.800	Seeds	3235	35	3341	3495	0.963	3416
NZA-37263	17.200	Seeds	3322	35	3440	3594	0.930	3509
NZA-33259	17.600	Seeds	3297	80	3331	3649	0.955	3487
NZA-37264	17.900	Seeds	2916	35	2875	3081	0.919	2998
NZA-33260	18.200	Seeds	10,830	240	11,994	13,145	1.000	12,672
NZA-34986	19.100	Bulk sediment	34,260	440	37,456	39,807	1.000	38,714
NZA-32805	20.000	Seeds	20,340	160	23,985	24,945	1.000	24,397

Core 2 (Horrocks et al., 2012b).

Lab code	Depth	Material	^{14}C yr BP	Error	LCR	UCR	RAPD	Median
NZA-37242	1.700	Seeds	−646[a]	30[a]	ND	ND	ND	ND
NZA-37243	2.800	Seeds	389	30	323	418	0.591	399
NZA-37214	3.400	Seeds	188	25	135	233	0.462	182
NZA-37248	4.300	Seeds	377	30	320	486	1.000	394
NZA-37247	5.000	Seeds	1025	30	800	935	0.988	864
NZA-37369	5.300	Seeds	1971	40	1806	1997	0.950	1879
NZA-32806	5.800	Seeds	1067	50	799	991	0.920	930

[a]Invalid age for calibration curve SHcal13.

Core 3 (Horrocks et al., 2012b).

Lab code	Depth	Material	^{14}C yr BP	Error	LCR	UCR	RAPD	Median
NZA-37251	3.600	Seeds/plant macroremains	572	30	507	560	0.968	538
NZA-37249	4.800	Seeds/plant macroremains	414	30	435	501	0.585	451
NZA-32813	5.100	Seeds	416	35	430	503	0.566	447

Core 4 (Horrocks et al., 2012b).

Lab code	Depth	Material	^{14}C yr BP	Error	LCR	UCR	RAPD	Median
NZA-37244	11.100	Seeds/plant macroremains	1685	30	1430	1607	1.000	1546
NZA-37246	11.300	Seeds	1971	30	1818	1934	0.923	1879
NZA-37245	11.500	Seeds/plant macroremains	1684	30	1429	1607	1.000	1545
NZA-37370	11.800	Seeds	1994	40	1823	2004	1.000	1906
NZA-33513	12.000	Seeds	3177	90	3137	3566	0.975	3341

Core 5 (Horrocks et al., 2012b).

Lab code	Depth	Material	^{14}C yr BP	Error	LCR	UCR	RAPD	Median
NZA-33973	5.950	Plant macroremains	386	40	319	492	1.000	400
NZA-37372	7.250	Seeds	5155	50	5726	5948	0.972	5834
NZA-32815	8.550	Plant macroremains	−297[a]	35[a]	ND	ND	ND	ND
NZA-37371	8.550	Seeds	2579	45	2458	2755	0.998	2608

[a]Invalid age for calibration curve SHcal13.

Core KAO2 (Flenley, 1996; Butler and Flenley, 2001, 2010; Butler et al., 2004).

Lab code	Depth	Material	^{14}C yr BP	Error	LCR	UCR	RAPD	Median
NZA-18033	0.038	Bulk sediment	312	40	278	459	0.964	370
NZA-17986	0.038	Plant macroremains	191	40	132	289	0.663	171
NZA-18034	0.038	Macroremains/pollen	824	40	662	764	1.000	706
NZA-27729	0.795	Plant macroremains	ND	ND	350	435	ND	393
NZA-15461	1.303	Macroremains/pollen	5753	62	6393	6662	0.964	6506
NZA-27750	1.765	Plant macroremains	ND	ND	660	676	ND	668
NZA-15462	2.200	Macroremains/pollen	1965	70	1705	2017	0.998	1869
NZA-27727	2.475	Plant macroremains	ND	ND	501	518	ND	510
NZA-27751	2.775	Plant macroremains	ND	ND	475	505	ND	490
NUTA-3515	2.900	Bulk sediment	1120	110	768	1187	0.960	991
NUTA-3011	11.400	Bulk sediment	890	110	632	959	0.970	776
NZA-27752	12.575	Plant macroremains	ND	ND	1825	1894	ND	1860
NZA-15463	13.413	Macroremains/pollen	2201	60	2012	2317	1.000	2165
NZA-27753	14.085	Plant macroremains	ND	ND	1887	1949	ND	1918
NUTA-3013	14.900	Bulk sediment	1630	130	1272	1751	0.978	1495

Lab code	Depth	Material	^{14}C yr BP	Error	LCR	UCR	RAPD	Median
NZA-27754	15.975	Plant macroremains	ND	ND	3721	3860	ND	3791
NZA-11603	16.800	Plant macroremains	4009	60	4230	4584	0.975	4426
NZA-11639	16.800	Plant macroremains	5716	60	6315	6570	0.920	6461
NZA-11638	16.800	Macroremains/pollen	5975	60	6633	6941	0.999	6762
NUTA-3516	17.900	Bulk sediment	9510	160	10,371	11,193	0.977	10,765
NZA-11604	18.300	Plant macroremains	8418	60	9246	9526	0.983	9391
NZA-11640	18.300	Plant macroremains	8498	60	9395	9543	0.903	9474
NZA-11641	18.300	Pollen extract	8477	60	9367	9535	0.860	9456
NZA-27755	19.115	Plant macroremains	ND	ND	9532	9551	ND	9542
NZA-27732	19.815	Plant macroremains	ND	ND	9917	10,184	ND	10,051
NZA-15464	20.375	Pollen extract	8726	65	9535	9889	1.000	9655
NUTA-3012	20.565	Bulk sediment	9130	170	9695	10,683	1.000	10,241

Core KAO3 (KAO05-3A) (Gossen, 2007, 2011).

Lab code	Depth	Material	^{14}C yr BP	Error	LCR	UCR	RAPD	Median
CAMS-132962	0.005	Seeds	45	ND	ND	ND	ND	ND
CAMS-132963	0.850	Seeds	245	40	138	230	0.582	203
CAMS-132964	1.200	Seeds	545	35	499	553	1.000	527
CAMS-132965	1.950	Seeds	605	30	520	567	0.654	555
CAMS-125890	12.650	Seeds	1200	35	965	1114	0.856	1051
CAMS-123153	12.750	Pollen extract	2480	60	2353	2711	1.000	2520
CAMS-123142	12.980	Seeds	1340	35	1172	1298	0.947	1225
CAMS-123143	13.310	Seeds	1320	30	1161	1278	0.881	1220
CAMS-125893	14.080	Seeds	1335	30	1176	1293	0.965	1225

Continued

Core KAO3 (KAO05-3A) (Gossen, 2007, 2011)—cont'd

Lab code	Depth	Material	^{14}C yr BP	Error	LCR	UCR	RAPD	Median
CAMS-123145	15.020	Seeds	1885	35	1705	1842	0.948	1781
CAMS-123154	15.020	Pollen extract	1910	130	1518	2148	1.000	1806
CAMS-125891	15.860	Seeds	2870	120	2743	3253	0.985	2968
CAMS-123155	15.970	Pollen extract	2890	130	2747	3273	0.964	2994
CAMS-135732	17.000	Seeds	3440	35	3559	3729	0.934	3644
CAMS-135734	17.000	Seeds	3435	40	3552	3729	0.900	3638
CAMS-135733	17.020	Seeds	6385	45	7169	7335	0.848	7277
CAMS-123147	17.040	Seeds	3470	70	3548	3868	0.952	3688
CAMS-123156	17.040	Pollen extract	3555	35	3689	3894	0.982	3781
CAMS-135731	17.080	Seeds	3185	40	3238	3452	1.000	3364
CAMS-123157	17.800	Pollen extract	7470	100	8027	8403	1.000	8245
CAMS-123219	17.810	Seeds	8370	320	8541	10,157	1.000	9273
CAMS-123148	18.920	Seeds	9210	100	10,170	10,593	0.992	10,359
CAMS-123158	18.920	Pollen extract	7730	210	8043	9014	1.000	8524
CAMS-123149	20.270	Seeds	9320	70	10,253	10,604	0.961	10,456
CAMS-123159	20.270	Pollen extract	8270	110	8976	9489	0.987	9204
CAMS-123160	20.720	Pollen extract	7610	200	7970	8789	0.968	8379
CAMS-123150	20.720	Seeds	9690	400	10,118	12,405	0.983	11,043
CAMS-123151	20.770	Seeds	11,030	130	12,696	13,087	1.000	12,879
NZA-23878	21.000	Seeds	11,017	45	12,724	12,980	1.000	12,825
CAMS-136170	21.500	Seeds	12,930	50	15,181	15,623	1.000	15,385

Core KAO08-03 (Rull et al., 2018; Seco et al., 2019).

Lab code	Depth	Material	^{14}C yr BP	Error	LCR	UCR	RAPD	Median
ULA–5790	0.106	Pollen extract	305	15	289	327	0.805	312
ULA–5792	0.125	Pollen extract	290	15	283	323	0.988	304
ULA–5789	0.141	Pollen extract	355	20	350	452	0.816	392
ULA–5791	0.160	Pollen extract	470	20	468	516	1.000	499
ULA–5872	0.196	Pollen extract	695	15	563	658	1.000	597
ULA–5821	0.216	Pollen extract	885	20	720	791	0.967	748
ULA–5874	0.247	Pollen extract	5060	20	5658	5891	1.000	5755
ULA–5817	0.271	Pollen extract	1060	20	906	962	0.977	934
ULA–5873	0.297	Pollen extract	1165	15	968	1059	1.000	1017

Rano Raraku

Core RRA3 (Flenley, 1979b; Flenley and King, 1984; Flenley et al., 1991).

Lab code	Depth	Material	^{14}C yr BP	Error	LCR	UCR	RAPD	Median
SRR–2425	0.950	ND	480	60	432	552	0.780	487
SRR–1553	1.200	ND	6850	50	7571	7751	1.000	7646
SRR–1679	1.700	ND	7770	90	8360	8768	1.000	8517
SRR–2426	2.150	ND	8060	70	8633	9035	0.972	8872
SRR–2427	3.700	ND	11,300	170	12,766	13,424	1.000	13,113
SRR–1554	4.750	ND	11,970	140	13,466	14,099	1.000	13,771
SRR–2428	6.775	ND	18,880	190	22,334	23,199	1.000	22,718
SRR–1555	7.750	ND	20,780	350	24,132	25,736	1.000	24,953
SRR–2429	10.900	ND	23,760	250	27,435	28,344	1.000	27,830
SRR–1352	12.400	ND	35,260	925	37,639	41,694	1.000	39,746

Core RRA4 (Flenley et al., 1991).

Lab code	Depth	Material	^{14}C yr BP	Error	LCR	UCR	RAPD	Median
Beta-10102	6.350	ND	7650	120	8162	8649	0.998	8413
Beta-10103	7.260	ND	9000	150	9583	10,415	0.996	10,030
Beta-10101	13.550	ND	13,770	480	15,220	17,891	1.000	16,582
SRR-2554	13.935	ND	13,810	140	16,218	17,081	1.000	16,646

Core SW (Dumont et al., 1998).

Lab code	Depth	Material	^{14}C yr BP	Error	LCR	UCR	RAPD	Median
ND	1.325	Plant macroremains	588	60	499	650	1.000	554
GIF A 92327	1.650	Bulk sediment	16,090	170	18,925	19,791	1.000	19,356
GIF A 92328	2.200	Bulk sediment	24,590	310	27,899	29,249	1.000	28,575

Core RRA5 (Azizi and Flenley, 2008).

Lab code	Depth	Material	^{14}C yr BP	Error	LCR	UCR	RAPD	Median
NZA-22678	6.880	Pollen extract	10,465	65	12,028	12,445	0.906	12,273
NZA-22831	9.700	Pollen extract	14,788	75	17,703	18,163	1.000	17,943
NZA-22832	11.650	Pollen extract	22,360	150	26,161	27,060	1.000	26,582
NZA-22833	13.250	Pollen extract	27,650	240	31,036	31,911	1.000	31,398

Core 1 (Mann et al., 2008).

Lab code	Depth	Material	^{14}C yr BP	Error	LCR	UCR	RAPD	Median
Beta–131055	0.135	Seeds	830	40	665	767	1.000	708
CAMS–78688	0.170	Pollen extract	3880	40	4145	4411	0.963	4249
CAMS–78689	0.203	Pollen extract	3850	40	4083	4359	0.942	4201
Beta–135105	0.340	Seeds	4380	50	4826	5050	0.989	4921
Beta–143721	0.755	Seeds	5200	50	5746	6005	0.987	5915
Beta–131056	0.965	Seeds	5590	40	6279	6414	0.995	6346

Core 2 (Mann et al., 2008).

Lab code	Depth	Material	^{14}C yr BP	Error	LCR	UCR	RAPD	Median
Beta–135106	0.145	Seeds	260	40	142	226	0.503	218
Beta–135107	0.175	Seeds	560	40	499	561	0.942	534
Beta–135108	0.245	Seeds	4320	40	4808	4970	0.867	4849
Beta–143722	0.505	Seeds	5210	50	5841	6018	0.762	5926
Beta–135109	0.940	Seeds	6770	50	7499	7675	1.000	7594

Core Lake (Horrocks et al., 2012a).

Lab code	Depth	Material	^{14}C yr BP	Error	LCR	UCR	RAPD	Median
NZA–37803	0.100	Seeds	1884	50	1696	1899	0.944	1780
NZA–37802	0.200	Seeds	1167	50	932	1112	0.926	1024
NZA–37804	0.400	Seeds	93[a]	45[a]	ND	ND	ND	ND
NZA–37005	0.600	Seeds	561	20	513	550	1.000	534
NZA–37006	0.800	Seeds	587	30	513	562	0.860	545
NZA–37008	0.900	Seeds	1881	25	1713	1831	1.000	1779
NZA–37014	1.000	Seeds	8938	35	9887	10191	0.989	10031
NZA–37010	1.700	Seeds	21,730	100	25,757	26,124	1.000	25,941
NZA–37939	2.000	Seeds	29,170	250	32,693	33,847	1.000	33,322
NZA–32811	2.250	Seeds	24,210	190	27,813	28,621	1.000	28,205

[a]Invalid age for calibration curve SHcal13.

Core RAR01 (Sáez et al., 2009).

Lab code	Depth	Material	^{14}C yr BP	Error	LCR	UCR	RAPD	Median
Poz–18688	ND	Pollen extract	13,950	70	16,550	17121	1.000	16,854
Poz–19930	0.370	Pollen extract	975	30	774	920	1.000	852

Core RAR02 (Sáez et al., 2009).

Lab code	Depth	Material	^{14}C yr BP	Error	LCR	UCR	RAPD	Median
Poz–19933	0.250	Pollen extract	795	35	650	736	0.999	690
Poz–19931	ND	Pollen extract	2580	40	2485	2753	0.993	2614
Poz–18743	ND	Pollen extract	18,180	120	21,651	22,329	1.000	21,993

Core RAR03 (Sáez et al., 2009; Cañellas-Boltà, 2014; Cañellas-Boltà et al., 2012, 2014, 2016).

Lab code	Depth	Material	^{14}C yr BP	Error	LCR	UCR	RAPD	Median
Poz-20530	0.170	Plant macrorests	109[a]	0.4[a]	ND	ND	ND	ND
Poz-19934	0.200	Pollen extract	3025	35	3004	3253	0.953	3155
Poz-24023	0.300	Plant macrorests	112[a]	0.4[a]	ND	ND	ND	ND
Poz-33774	0.540	Pollen extract	4080	40	4416	4630	0.921	4514
Poz-24024	0.800	Plant macrorests	100[a]	0.4[a]	ND	ND	ND	ND
Poz-33775	1.280	Plant macrorests	4670	40	5283	5473	0.944	5397
Poz-24025	1.550	Plant macrorests	490	35	455	541	1.000	506
Poz-20571	1.850	Plant macrorests	5030	40	5606	5770	0.767	5717
Poz-19935	2.300	Pollen extract	5450	40	6173	6297	0.816	6222
Poz-24026	2.850	Plant macrorests	3640	35	3825	3993	0.926	3905
Poz-24027	3.550	Plant macrorests	6170	40	6898	7160	1.000	7020
Poz-24030	4.140	Plant macrorests	6620	50	7420	7573	1.000	7482
Poz-18689	4.650	Pollen extract	6960	40	7664	7851	0.999	7748
Poz-24031	5.340	Plant macrorests	7410	50	8038	8326	1.000	8189
Poz-24032	6.150	Plant macrorests	7930	50	8582	8814	0.759	8712
Poz-18690	6.830	Pollen extract	8010	40	8691	8994	0.939	8841
Poz-18691	7.330	Pollen extract	8340	50	9128	9453	1.000	9309
Poz-19936	8.350	Pollen extract	9810	60	11,075	11,312	0.942	11,195
Poz-18693	10.390	Pollen extract	10,430	50	12,026	12,425	1.000	12,230
Poz-18694	11.250	Pollen extract	11,020	50	12,724	12,989	1.000	12,834
Poz-18696	13.390	Pollen extract	13,570	70	16,047	16,560	1.000	16,291
Poz-18695	13.590	Pollen extract	14,010	70	16,622	17,209	1.000	16,949

[a]Invalid ages for calibration curve SHcal13.

Core RAR04 (Sáez et al., 2009).

Lab code	Depth	Material	^{14}C yr BP	Error	LCR	UCR	RAPD	Median
Poz–18697	ND	Pollen extract	4535	35	5035	5301	0.950	5162

Core RAR05 (Sáez et al., 2009).

Lab code	Depth	Material	^{14}C yr BP	Error	LCR	UCR	RAPD	Median
Poz–19937	ND	Pollen extract	103[a]	0.4[a]	ND	ND	ND	ND
Poz–18699	ND	Pollen extract	5660	40	6306	6484	1.000	6393

[a]Invalid ages for calibration curve SHcal13.

Core RAR07 (Sáez et al., 2009; Cañellas-Bolta, 2014; Cañellas-Boltà et al., 2012, 2014, 2016).

Lab code	Depth	Material	^{14}C yr BP	Error	LCR	UCR	RAPD	Median
Poz–18700	3.110	Pollen extract	7940	50	8591	8814	0.721	8727
Poz–18701	3.330	Pollen extract	8680	40	9531	9693	1.000	9594
Poz–18703	7.250	Pollen extract	13,010	60	15,266	15,748	1.000	15,503
Poz–18704	7.450	Pollen extract	13,500	50	15,983	16,405	1.000	16,193
Poz–19938	9.230	Pollen extract	18,850	130	22,409	22,989	1.000	22,676
Poz–19939	10.920	Pollen extract	24,340	230	27,845	28,771	1.000	28,324
Poz–18705	11.540	Pollen extract	30,060	240	33,698	34,547	1.000	34,088

Core RAR08 (Sáez et al., 2009; Cañellas-Boltà, 2014; Cañellas-Boltà et al., 2013).

Lab code	Depth	Material	^{14}C yr BP	Error	LCR	UCR	RAPD	Median
Poz–42955	0.040	Pollen extract	106.94[a]	0.39[a]	ND	ND	ND	ND
Poz–42957	0.090	Pollen extract	101.03[a]	0.35[a]	ND	ND	ND	ND
Beta–316585	0.110	Pollen extract	380	30	322	486	1.000	395
Poz–32007	0.130	Pollen extract	505	30	490	541	1.000	512
Poz–42958	0.170	Pollen extract	840	30	671	754	0.989	710
Beta–316586	0.180	Pollen extract	780	30	649	728	0.982	677
Beta–316587	0.190	Pollen extract	1180	30	959	1093	0.983	1021
Poz–42959	0.210	Pollen extract	2120	35	1992	2152	0.941	2053
Beta–316588	0.230	Pollen extract	2970	30	2958	3179	1.000	3075
Poz–19940	0.250	Pollen extract	2160	30	2007	2159	0.923	2100
Poz–42960	0.290	Pollen extract	4800	35	5448	5589	0.831	5511
Poz–32120	0.410	Pollen extract	4530	45	4971	5306	1.000	5151
Poz–18706	ND	Pollen extract	5910	40	6558	6790	1.000	6689

[a]Postmodern carbon.

CHAPTER 4

Paleoecological pioneers: The rising of the ecocidal paradigm

Contents

4.1	Before paleoecology	138
4.2	The first systematic pollen analyses	139
4.3	The first paleoecological synthesis	140
	4.3.1 Vegetation dynamics	141
	4.3.2 Paleoclimatic and paleoenvironmental inferences	149
	4.3.3 Human impact	151
	4.3.4 General conclusions	153
4.4	The first socioecological synthesis	153
4.5	Some insights on the pioneering works	158
	4.5.1 Deforestation chronology	159
	4.5.2 Paleoclimatic inference and paleoecological implications	160
	4.5.3 The upper altitudinal forest limit	161
	4.5.4 Human and climatic impact	162
	4.5.5 The socioecological catastrophe	163
Appendix 4.1		163

This chapter summarizes the pioneering paleoecological research developed on Easter Island, with emphasis on the work by John Flenley (to whom this book is dedicated) and his colleagues in the late 20th century, mostly during the 1980s. The first part explains the origin of the initial interest and the first steps toward the paleoecological study of Easter Island's paleoecology, to introduce the seminal publications of Flenley and his team. The results and interpretations of these authors are presented in the second part, which is subdivided into two main sections. The first section is a short account of the initial high-impact publications that brought Easter Island to the attention of the general scientific community and the society (Flenley and King, 1984; Dransfield et al., 1984). The second section is more extended and provides more details about vegetation dynamics, environmental reconstruction, and human impact (Flenley et al., 1991). Both sections summarize the results obtained by Flenley and coworkers in its original form, without any additional comments. At the end of the chapter, some personal

Paleoecological Research on Easter Island
https://doi.org/10.1016/B978-0-12-822727-5.00004-0

reflections on this pioneering work are added to introduce a number of relevant points that remained unclear and boosted further research, as discussed in the following chapters.

4.1 Before paleoecology

To understand the origin of paleoecological studies on Easter Island, it is necessary to come back to the 1955–56 Norwegian expedition led by Thor Heyerdahl, who was the first to retrieve some short sediment cores in Rano Kao and Rano Raraku. The Swedish palynologist Olof Selling, with wide experience in Hawaiian paleoecology (Selling, 1946–1948), analyzed the pollen content of these undated cores and noted the common presence of pollen from some Compositae and a palm tree species that was already extinct on the island and that the author assigned provisionally to some *Pritchardia* species. Selling concluded that the island had once born a forest. Unfortunately, this was never published and the samples were never reanalyzed; the only mention to this finding is in the form of a personal communication from Selling to Heyerdahl (Heyerdahl and Ferdon, 1961). The first European visitors reported that the island was unforested; therefore, if Selling's forests truly existed, they would have disappeared before 1722. This added another enigma to the Easter Island's prehistory. Interestingly, it was also Heyerdahl who, in 1958, collected seeds of the last Easter Island's wild toromiro (*Sophora toromiro*) specimen, which were planted at the Göteborg Botanical Garden, where its descendants are still cultivated and have been the source of other specimens growing in other botanical gardens around the world (Maunder et al., 2000)—although Shepherd et al. (2020) suggested that some European specimens could have been collected by James Cook in the late 18th century (Section 2.1). This is why this species has not yet become extinct. It could be said that, despite his low popularity within the scientific community for his seemingly awkward theories (Flenley and Bahn, 2003), Heyerdahl planted the seeds not only to allow the survival of the toromiro but also to launch the paleoecological study of Easter Island.

The idea of a forested island, as inferred from Selling's palynological analyses on Heyerdahl's sediment cores, was taken seriously by William Mulloy, who was also part of Heyerdahl's expedition (Section 2.1). This researcher believed that tree trunks were needed to transport the moai and to emplace them in their stone platforms across the island (Section 2.4.2). He wrote: "…statues were transported face down on prepared roads probably on wooden sledges pulled by many men …" (Mulloy, 1974). Therefore, the

deforestation of the island would have caused the end of the moai industry, tentatively dated back to 1680 CE by that time (Mulloy, 1970a). In his
excavations, Mulloy found root molds that he attributed to the past occurrence of a vegetation significantly different than the present. The further
finding of carbonized wood remains was interpreted as evidence of former
forested vegetation and was used to relate forest removal with human crisis, social conflicts, and cultural decay of the prehistoric Rapanui society
(Mulloy, 1979). According to Flenley and Bahn (2003), this was "...the seeds
of Easter Island being considered a microcosm of, and a warning about, the
current planetary situation, although Mulloy did not have any idea of the
scale of the island's deforestation or any precise data about what constituted
the forests." In a paper written in 1976 but not published until two decades
later, Mulloy (1997) emphasized the need for repeating and extending the
paleobotanical studies initiated by Selling to test the hypothesis of a formerly forested island. The British palynologist John Flenley picked up the
gauntlet and initiated the systematic paleoecological study of Easter Island,
as explained in detail in this chapter.

4.2 The first systematic pollen analyses

Between 1977 and 1983, Flenley and his team retrieved several cores from
Rano Raraku, Rano Aroi, and Rano Kao using a Russian borer (Jowsey,
1966). The first stratigraphic observations on these cores suggested the occurrence of two possible ecological changes, the first probably related with
a climatic shift from drier to wetter conditions and the second consisting of
a major decline in vegetation cover. However, pollen analysis and radiocarbon dating were still in progress and these inferences were considered preliminary (Flenley, 1979b). The first dated palynological reconstructions were
published five years later by Flenley and King (1984) who suggested that the
island would have been covered by palm-dominated forests during the last
37,000 years and that these forests disappeared during the last millennium
probably due to anthropogenic causes, which could have led to a cultural
collapse of the prehistoric Rapanui culture. These authors also concluded that
climate was cooler and/or drier than at present between 21,000 and 12,000
^{14}C yr BP, a time interval that includes the Last Glacial Maximum (LGM).

The identity of the palm that dominated Easter Island's forests, however,
was—and still is—a mystery. As mentioned in Section 1.4.4, the pollen was
similar to several palm genera, including *Cocos* and *Pritchardia*, which are
widespread in the Pacific island, or *Jubaea chilensis*, common in the Pacific

coasts of Chile. Of them, only *Cocos nucifera* occurs today on the island but as a result of recent introductions. Dransfield et al. (1984) reported the finding of several palm endocarps in a cave near Ana Okeke, in the Poike Peninsula. These endocarps were of similar size but different morphology to those of *J. chilensis* and the authors concluded that they were from a palm related but distinct to this species. Radiocarbon dating on several of these palm fruits gave ages of 820 ± 40 ^{14}C yr BP, which coincided with the time of deforestation, i.e., the last millennium, and also with the flourishment of the moai industry (Dransfield et al., 1984). The endocarps showed clear signs of rodent gnawing that these authors attributed to *Rattus concolor*, a Polynesian rat probably introduced by the first settlers. Similar endocarps were found on other archaeological excavations suggesting that the palm fruits would have also been part of the human diet. Dransfield et al. (1984) concluded that the extinction of the unknown palm could have been due to "…a combination of direct deforestation and prevention of reproduction, the latter resulting from eating of the fruits by man and the introduced rats." These authors also mentioned that the extinct palm, if similar to *J. chilensis*, would have been used in the transportation and final emplacement of the moai in their respective ahu, as initially proposed by Mulloy (1970a). Based on the same endocarps, Dransfield (in Zizka, 1991) decided to provisionally describe a new genus and species of extinct palm, named *Paschalococos disperta*, until more material, especially floral details, were available for comparison with known extant genera.

4.3 The first paleoecological synthesis

Further studies on the same cores of Flenley and King (1984) including more radiocarbon dates, and multiproxy biological (pollen) and physicochemical analyses led to the first island-wide paleoecological synthesis (Flenley et al., 1991). These authors already recognized the impossibility of developing modern-analog studies due to the total destruction of the original vegetation (Section 1.4.4). Therefore, they based the interpretation of their pollen diagrams in terms of vegetation development using an intuitive approach based on the indicator potential of pollen types identified. A summary of the pollen types found, their potential parent plants and their corresponding indicator power is provided in Appendix 4.1. The main results of this seminal paleoecological study are presented in the following section with some detail, not only for their pioneering character but also because these findings already put on the table many ideas that have been tested, developed, and/or discussed in further works.

4.3.1 Vegetation dynamics

The pollen record from Rano Raraku of Flenley et al. (1991) corresponded to core RRA3, situated in the eastern margin, just in front of the inner crater wall with maximum human activity (moai carving) during prehistoric times (Fig. 3.6). The age-depth model assumed more or less constant sediment accumulation rates throughout the core, which represented the last 35,000 ^{14}C yr BP (Fig. 4.1). The authors did not indicate how this linear model was obtained but mentioned that three dates situated around 2 m

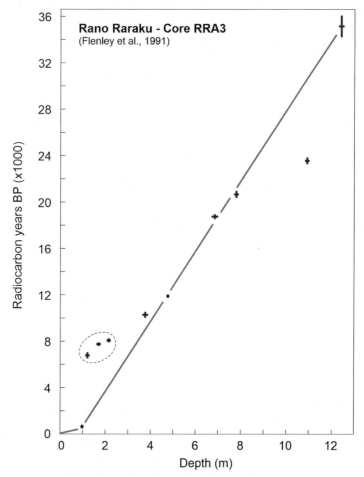

Fig. 4.1 Age-depth model used by Flenley et al. (1991) in Rano Raraku core RRA3. The horizontal *red* bars represent the length of the samples dated and vertical *red* bars are the dating error. The *blue* line is the linear model used by Flenley et al. (1991) and the ages encircled by a dotted line are the rejected dates. The *green* line represents the accumulation rate of the upper 70 cm of the core after ^{210}Pb analysis.

depth gave ages "anomalously old" and suggested that these samples were contaminated by older soil or bedrock carbon (Fig. 4.1). Radiocarbon dating was complemented with ^{210}Pb analysis of the core top, which revealed accumulation rates of 1 cm/year thus indicating that the uppermost 70 cm were accumulated in only 70 years.

The continuous presence of fine organic detritus (gyttja) suggested that the Raraku crater was permanently occupied by a lake. Lake levels may have been higher than at present during some phases, as evidenced by the presence of dried gyttja above the present water level, at some points around the lake. The main vegetation changes were reconstructed by pollen analysis (Fig. 4.2). The present belt and overgrowth of subaquatic macrophytes (hydrosere) (Figs. 1.6 and 3.6) seems to be relatively recent (Early-Mid Holocene), as the dominant species (*Scirpus californicus*) did not fully establish until about 8000 ^{14}C yr BP and another important component, *Polygonum acuminatum*, was absent or very scarce before 500 ^{14}C yr BP. Regarding the catchment vegetation, pollen analyses revealed the occurrence of four main phases of development (Fig. 4.2). Between 35,200 and 28,000 ^{14}C yr BP (Zone I), a palm-dominated forest covered the site. A shrub layer composed of *Sophora*, Compositae-Tubuliflorae, *Triumfetta*, Urticaceae/Moraceae, and *Macaranga* probably existed and a ground flora of ferns and grasses was also likely present. *Scirpus* was already present but in low amounts. During the second phase (Zone II: 28,000–12,000 ^{14}C BP), the forest may have been more open thus facilitating the development of a more dense and diverse shrub layer where *Sophora* was much more abundant than in the previous phase and new elements such as *Coprosma*, *Trema*, *Sapindus*, and Myrtaceae appeared. Grasses were also more abundant than in Zone I.

The situation reverted between 12,000 and 1200 ^{14}C yr BP (Zone III), and the forest returned to a situation similar to Zone I, with the exception of *Triumfetta*, which was more abundant. Ferns were also more abundant and the marginal hydrosere, dominated by *Scirpus* and *Cyperus*, was in full development. The start of Zone IV (1200 ^{14}C yr to present) was marked by a sharp decline in the pollen of Palmae and the elements of the shrub layer, which were replaced by grasses and other herbs (likely introduced) such as *Plantago*, Caryophyllaceae, Compositae-Liguliflorae, *Rumex,* and *Euphorbia*. The hydrosere attained its maximum development. The dramatic changes observed during Zone IV were considered to be consistent with the destruction of the native forest by humans, presumably by a combination of felling and burning (although not counted, charcoal particles were abundant in this pollen zone). As previously suggested by Dransfield et al. (1984), the introduction of rats may have also contributed to palm decline. Just after palm

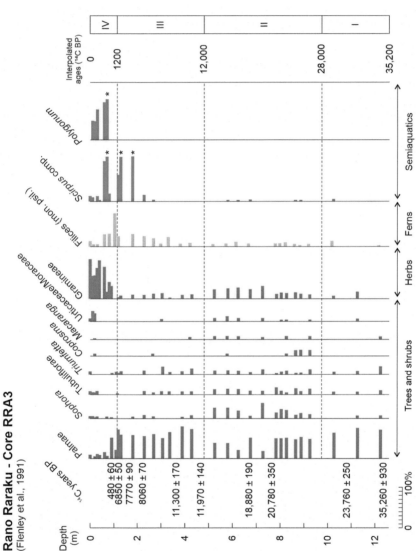

Fig. 4.2 Simplified percentage pollen diagram of core RRA3 with the most important taxa used by Flenley et al. (1991) to define the pollen zones (I to IV) and to reconstruct the Late Pleistocene and Holocene vegetation dynamics in the Raraku catchment. Semiaquatics are outside the pollen sum on which the percentage was calculated and values over 100% are indicated by an asterisk.

decline, ferns increased due to light availability on the ground layers and were progressively replaced by herbs introduced by humans for cultivation and grazing. The development of the hydrosere may have been facilitated by the enhanced nutrient supply due to human activities around the lake. The apparent recovery of trees at the top of the zone could be due to the reworking of older marginal sediments into the center of the lake. By the time of development of these pioneering pollen analyses, the accepted date for initial human settlement of Easter Island was 400 CE (Heyerdahl and Ferdon, 1961), which was used by Flenley et al. (1991) to support the hypothesis of humans as the main responsible for forest clearing.

The Kao pollen record of Flenley et al. (1991) was obtained in core KAO1, situated in the NW margin of the lake (Fig. 3.6). According to the age–depth model (also obtained with unspecified methods), this record encompassed only the last 1500 years. This date was extrapolated by the authors from the trends of the upper four radiocarbon dates, as the basal age of the diagram was really 990 ± 70 ^{14}C yr BP (Fig. 4.3). In this case, charcoal particles were present along the whole core thus suggesting continuous human disturbance by fire (Fig. 4.4). Three main vegetation phases (pollen zones) were recognized in this core. Between 1550 and 1300 ^{14}C yr BP (Zone I), the Kao catchment was still forested, with *Triumfetta*, the unknown palm and *Sophora* as the main elements. The authors suggested that the forest may have been recovering from a previous disturbance, as *Triumfetta*, an element from the shrubby understory, progressively gave way to the taller palms. The marginal swamp was probably dominated by *Polygonum* and *Scirpus* was scarcer. The destruction of most Kao forests took place between 1300 and 950 ^{14}C yr BP (Zone II), starting with the decline of *Triumfetta*, followed by *Sophora* and the palms. The final demise of forests occurred in Zone III (950 ^{14}C yr BP to present), with the disappearance of palms and *Triumfetta* (*Sophora* remained until the present). Grasses dominated the vegetation and some new taxa appeared that were considered human introductions (*Melia, Casuarina, Trema, Psidium*). The presence of *Apium* suggested its cultivation at the lake margin. The hydrosere, dominated by *Polygonum* and *Scirpus*, experienced a significant growth, possibly as a result of increased nutrient supply due to forest clearance. Another core (KAO2) retrieved in Rano Kao by Flenley and his coworkers in 1983, in a slightly more central position (Fig. 3.6), was analyzed later (Flenley, 1996; Butler and Flenley, 2010) and will be discussed in Chapters 5 and 6.

The Rano Aroi pollen record of Flenley et al. (1991) proceeded from core ARO1, situated near the center of the swamp (Fig. 3.6). According to the age-depth model, this record accounted for the last 38,000 years (Fig. 4.5). In this case, the authors did not attempt to establish a single

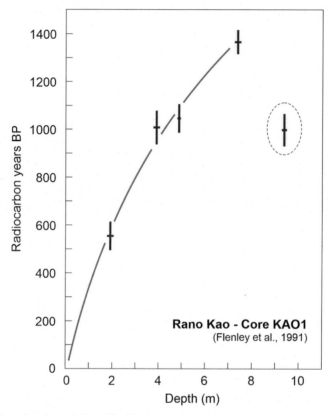

Fig. 4.3 Age-depth model used by Flenley et al. (1991) in Rano Kao core KAO1. The horizontal *red* bars represent the length of the samples dated and vertical *red* bars are the dating error. The *blue* line is the model used by Flenley et al. (1991) and the age encircled by a dotted line is the rejected date.

age–depth model for the whole sequence but used local interpolations to obtain three different linear trends representing variable accumulation rates (Fig. 4.5). The top sample was excluded from the model, as its age (almost 19,000 ^{14}C yr BP) was considered anomalous as a consequence of vertical mixing due to human reworking of sediments (Section 3.5). The authors suggested that the top meter of this section should be disregarded for dating and stratigraphic purposes, as it probably consisted of a mixture of material of different ages. They also mentioned the possibility of contamination of sediments older than 20,000 years with younger plant roots penetrating the peat deposit. Therefore, the base of the sequence could be much older than the dates indicated.

The Aroi record was subdivided into seven pollen zones (I to VII), excluding the uppermost meter, which was considered too disturbed for a

Rano Kao - Core KAO1
(Flenley et al., 1991)

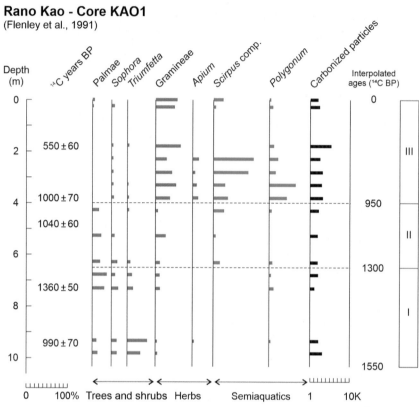

Fig. 4.4 Simplified percentage pollen diagram of core KAO1 with the most important taxa used by Flenley et al. (1991) to define the pollen zones (I to III) and to reconstruct the Late Holocene vegetation dynamics in the Kao catchment. Carbonized particles are represented as counts using a logarithmic scale.

reliable paleoecological reconstruction (Fig. 4.6). Between 38,000 and 34,900 ^{14}C yr BP (Zone I), the site was covered by grass meadows and the low abundance of ferns and aquatic elements were interpreted by Flenley et al. (1991) in terms of a dry peat surface with the water table much lower than today. The moderate values of palms, as compared with Rano Kao and Rano Raraku (Figs. 4.2 and 4.4), suggested that the unknown palm species was mainly a lowland element and its pollen would have reached the Aroi swamp (430 m elevation) by uphill wind dispersal. The vegetation around the site probably consisted of grasses and shrubs (*Coprosma* and Compositae-Tubuliflorae). The dryness accentuated between 34,900 and 33,200 ^{14}C yr BP (Zone II), as indicated by the absence of ferns and aquatic plants, along with the general absence of pollen from some horizons, which suggested aerial exposure and oxidation of the peat surface. Between 33,200

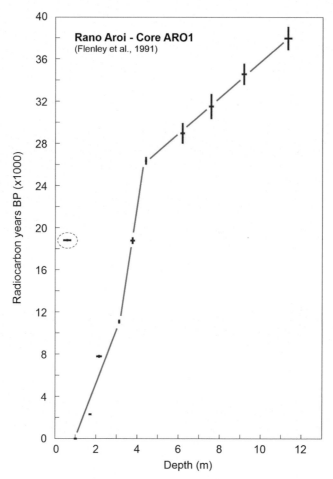

Fig. 4.5 Age-depth model used by Flenley et al. (1991) in Rano Aroi core ARO1. The horizontal *red* bars represent the length of the samples dated and vertical *red* bars are the dating error. The *blue* line is the model used by Flenley et al. (1991) and the age encircled by a dotted line is the rejected date.

and 27,950 ^{14}C yr BP (Zone III), the swamp was dominated by *Scirpus* and some ferns, whereas the surrounding dryland was covered mostly by Tubuliflorae shrublands, with grasses less abundant than formerly in both environments. The upper forest limit would have been above the coring site. The situation changed in Zone IV when grass meadows recovered their former importance for a while (27,950–26,500 ^{14}C yr BP) and the upper forest limit shifted slightly below the Aroi swamp. The Tubuliflorae shrublands recovered their former importance between 26,500 and 16,600 ^{14}C yr BP (Zone V), perhaps by the ascent of the upper forest limit, and the swamp was strongly dominated by *Scirpus*. Between 16,600 and 12,000

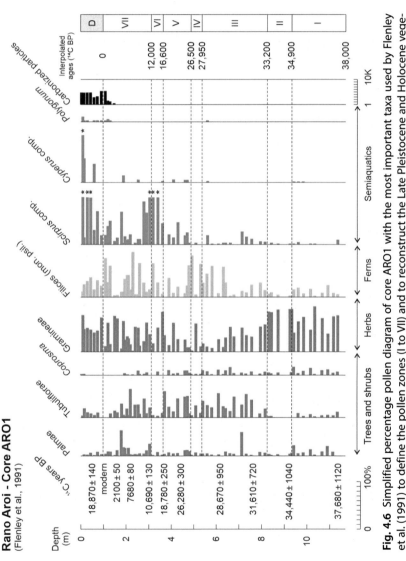

Fig. 4.6 Simplified percentage pollen diagram of core ARO1 with the most important taxa used by Flenley et al. (1991) to define the pollen zones (I to VII) and to reconstruct the Late Pleistocene and Holocene vegetation dynamics in the Aroi catchment. Carbonized particles are represented as counts using a logarithmic scale. Semiaquatics are outside the pollen sum on which the percentage was calculated and values over 100% are indicated by an asterisk. D, disturbed.

^{14}C yr BP (Zone VI), *Scirpus* started to totally dominate the swamp and the surrounding vegetation was less shrubby, probably by another depression of the upper forest limit. Finally, the last 12,000 years (Zone VII) have been characterized by a new climbing of the upper forest limit above the site and the recovery of the Tubuliflorae-*Coprosma* scrub around the swamp, which remained dominated by *Scirpus*. In the uppermost levels of Zone VII, woody plants started to disappear and carbonized particles appeared for the first time, suggesting anthropogenic burning. Although the uppermost meter had been previously considered useless for paleoecological reconstruction by Flenley et al. (1991), these authors suggested that this interval could represent the 20th century and highlighted the disappearance of palms and the appearance of introduced plants such as *Plantago, Eucalyptus,* and probably some Tubuliflorae weeds.

4.3.2 Paleoclimatic and paleoenvironmental inferences

Flenley et al. (1991) inferred paleoclimatic changes from pollen data using a temperature index based on the ratio between Palmae and Compositae-Tubuliflorae pollen. According to these authors, palms could be considered as indicators of warm climates because of their preference for tropical/subtropical lowlands. On the other hand, the arborescent Compositae (Tubuliflorae) of the Pacific islands might be considered as indicators of cool climates because they are mostly restricted to montane forests. Therefore, a ratio between these two pollen elements might be considered a warm/cool index. Similarly, the ratio between Gramineae (as representative of drier climates) and ferns (as typical of moist climates) was considered to be an index of moisture availability. These indices were calculated for Rano Raraku and Rano Aroi sequences discussed above to reconstruct the climatic trends during the last 38,000 years (Fig. 4.7). In addition to these climate changes, geochemical analyses provided complementary information for paleoenvironmental reconstruction.

In Rano Raraku, changes in organic matter content, mobile cations (Ca, Na, Mg), and bedrock oxides (Si, Al, Fe) suggested the occurrence of a phase of relatively high soil erosion until about 10,000 ^{14}C yr BP (near the base of Pollen Zone III), followed by a long period of reduced erosion and higher lake levels, probably as a consequence of a precipitation increase. During the last 1500 years (top of Pollen Zone III), an abrupt rise of Si, Fe, and Al, and the sharp drop of mobile cations and organic content coincided with the forest removal and the rapid decrease of the vegetation cover, which was attributed to human disturbance (Section 4.3.1). The Rano Aroi sediments were mostly organic and autochthonous but an increase of Al and

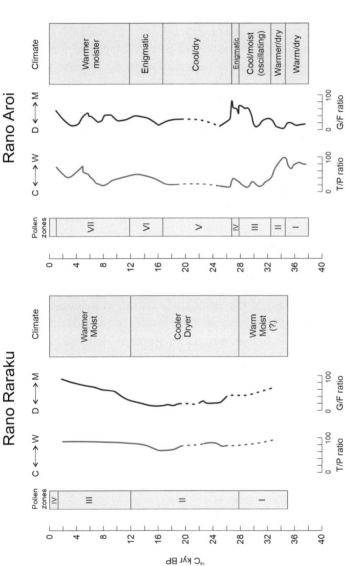

Fig. 4.7 Paleoclimatic interpretation of Rano Raraku and Rano Aroi sequences (Figs. 4.2 and 4.6) using the pollen ratios Tubuliflorae/Palms (T/P) as a temperature proxy, and Grasses/Ferns (G/F) as a moisture proxy. Pollen zones (I to IV for Raraku and I to VII for Aroi) are also included for comparison. C, cool; W, warm; D, dry; M, moist. (Redrawn and modified from Flenley, J.R., King, A.S.M., Jackson, J., Chew, C., Teller, J.T., Prentice, M.E., 1991. *The Late Quaternary vegetational and climatic history of Easter Island. J. Quat. Sci. 6, 85–115.*)

Fe at about 9.5 m (near the transition from pollen zones I and II; 34,900 ^{14}C yr BP) suggested the drying of the swamp, which coincided with the occurrence of dry grass-dominated vegetation in Zone II. During the last 2000 years (top of Pollen Zone VII), most cations showed high values suggesting catchment instability due to human activities. The same pattern was observed in Rano Kao, where almost all cations increased during the last millennium (Pollen Zone III) likely due to increased soil erosion and leaching, probably as a consequence of human disturbance of vegetation.

Combining the Raraku and Aroi sequences (Fig. 4.7), Flenley et al. (1991) emphasized that the Easter Island's climate was cooler and drier than the present between about 26,000 and 12,000 ^{14}C yr BP, which included the LGM interval. However, this cooling was insufficient to drive the vegetation around Rano Aroi, which was close to the upper forest limit, down 350 m to Rano Raraku, whose catchment remained fully forested. Using two extreme adiabatic lapse rates for the region (−0.5°C and −0.8°C per 100 m), Flenley et al. (1991) estimated that the temperature decrease needed for a 350-m downward vegetation sift is between 1.8°C and 2.8°C. Therefore, the Easter Island LGM cooling should have been less pronounced, which was in agreement with SE Pacific SST reconstructions of the time showing decreases up to 2°C during the LGM (e.g., Gates, 1976; Rind and Peteet, 1985).

4.3.3 Human impact

Flenley et al. (1991) highlighted that the expansion of grass meadows at the expense of the former forests recorded in the upper part of all pollen diagrams was coeval with a drop of organic matter, an abrupt increase of mineral inwash (especially metallic ions) to the waterbodies and the appearance, for the first time, of charcoal fragments. These authors also emphasized that there were no indications of climatic change associated with the forest decline and that Easter Island's forests had survived all climatic changes occurred during the last 30,000 years. Therefore, they concluded that human deforestation was the most likely cause for the observed vegetation change. Forest clearing would have started between 1200 and 800 ^{14}C yr BP in Rano Raraku and Rano Kao, and about 1000 ^{14}C yr BP in Rano Aroi. The total disappearance of forests would have occurred at about 500 ^{14}C yr BP, with Rano Kao as the last forested catchment. According to the archaeological chronology of the time, the collapse of the megalithic culture was dated back to 1680 CE (Heyerdahl and Ferdon, 1961)—not too far from 500 ^{14}C yr BP, according to Flenley et al. (1991). These authors concluded that Easter Island's deforestation may have led to an ecological disaster, as illustrated in Fig. 4.8. The starting point is human immigration

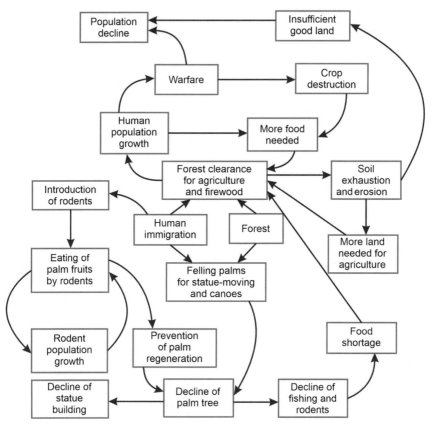

Fig. 4.8 Hypothetical cause-effect model for socioecological events on Easter Island. *(Redrawn and modified from Flenley, J.R., King, A.S.M., Jackson, J., Chew, C., Teller, J.T., Prentice, M.E., 1991. The Late Quaternary vegetational and climatic history of Easter Island. J. Quat. Sci. 6, 85–115.)*

and settlement, which triggered forest clearance for agriculture and fire-wood and resulted in human population growth. The introduction of rats that ate palm fruits prevented the regeneration of palm forests and, together with continued wood exploitation for moving the statues and building canoes, accelerated forest decline and food shortage, a decline in statue-carving, warfare, and human population crash.

Another aspect of human disturbance is the introduction of exotic plants. The study of Flenley et al. (1991) confirmed the postsettlement appearance of a number of taxa but also demonstrated that some species that were formerly considered to be introduced by humans were really native to the island. This was the case of *Triumfetta, Solanum, Euphorbia serpens*, and *Scirpus*, which was already present on the island 36,000 years ago.

4.3.4 General conclusions

The synthesis of all the results discussed above constituted the first general paleoenvironmental and paleoecological reconstruction for Easter Island, which was expressed by Flenley et al. (1991) into five main conclusions (Figs. 4.9 and 4.10):

1. Easter island was formerly forested, the main trees including an unknown palm, *S. toromiro* and *Triumfetta semitriloba*.
2. The uppermost altitudinal limit of this forest probably consisted of a shrub belt dominated by *Coprosma* and an unknown Compositae-Tubuliflorae species. This upper boundary was located near the elevation of Rano Aroi (430 m) but oscillated in response to climatic changes.
3. The climate of Easter Island was fluctuating between 38,000 and 26,000 ^{14}C yr BP. Between 26,000 and 12,000 ^{14}C yr BP, the climate was probably cooler and drier than the present. This cooling would have been of 2°C or less, as compared to present average temperatures. A precipitation (or moisture balance) increase occurred from 10,000 ^{14}C yr BP. The uppermost forest limit increased in elevation, which suggested a coeval rise in temperature.
4. Anthropogenic forest clearance took place between 1200 and 800 ^{14}C yr BP and the total forest demise occurred at 500 ^{14}C yr BP, which is compatible with the hypothesis that the decline of the megalithic culture was associated with deforestation.
5. The hypothesis that Easter Island is poor in woody species due to its isolation is not supported by these results, which are more compatible with the view that human activities are responsible for present-day floristic depauperation (see Section 1.4.2).

4.4 The first socioecological synthesis

Just a year after the pioneering paleoecological work of Flenley et al. (1991), Bahn and Flenley (1992) did the first socioecological synthesis by merging the recently obtained paleoecological evidence with the available archaeological, ethnographical, anthropological, and historical information. The concluding chapter of Bahn and Flenley's (1992) book was entitled: "The island that self-destructed" and suggested that, besides its evident religious motivations, the moai industry was also a strong competitive activity among tribes and clans to achieve the largest and most spectacular monuments. The moai workers were dedicated exclusively to this task and should be maintained by other social sectors dedicated to food production. As the moai activity increased, food producers had to support ever-increasing numbers

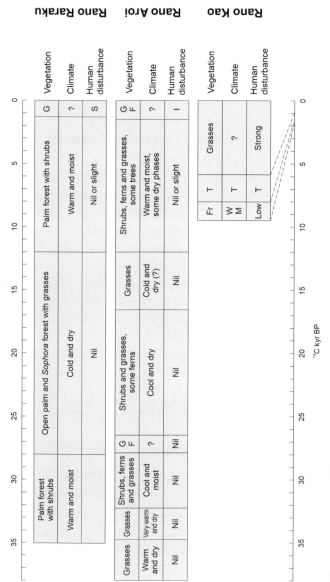

Fig. 4.9 Summary of the general conclusions of Flenley et al. (1991) regarding vegetation, climate and human disturbance in the three coring sites studied. G, grasses; F, ferns; Fr, ferns; T, forest; I, increasing; M, moist; S, sligth; T, transition; W, warm. *(Redrawn and modified from Flenley, J.R., King, A.S.M., Jackson, J., Chew, C., Teller, J.T., Prentice, M.E., 1991. The Late Quaternary vegetational and climatic history of Easter Island. J. Quat. Sci. 6, 85–115.)*

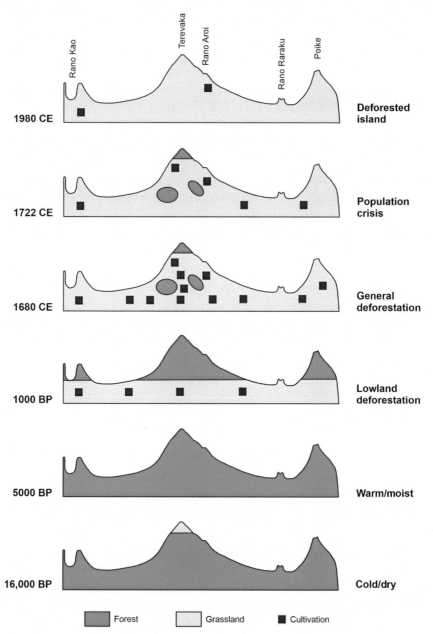

Fig. 4.10 Graphical synthesis of the main climatic, vegetation, and human trends and events since 16,000 ^{14}C BP, according to the paleoecological analyses of Flenley et al. (1991). *(Redrawn and modified from Bahn, P., Flenley, J., 1992. Easter Island, Earth Island. Tames & Hudson, London and Flenley, J.R., Bahn, P.G., 2003. The Enigmas of Easter Island. Oxford Univ. Press Oxford.)*

of nonfood producers (i.e., carvers), which led to deforestation and vegetation depletion by burning and cutting, leaching and soil erosion, increasing evaporation and, finally, a reduction in crop yields (Fig. 4.11). This would have also caused the drying up of springs and streams. At the same time, the disappearance of large timber for canoe building eventually led to the abandonment of deep-sea fishing and prevented the islanders to navigate to other islands. Bahn and Flenley (1992) did not dismiss the possibility of a drought that made things worst but they claimed that "…clearly the islanders brought disaster upon themselves by gradually destroying a crucial resource, the palm, and unwittingly preventing its regeneration through having imported rats." These authors agreed with Mulloy (1974) who considered the communal compulsion of the ancient Rapanui for giant statues and platforms "insane," to the point that basic subsistence activities such as farming and fishing were neglected.

The argument for dismissing a potential dominant effect of an eventual drought on deforestation has already been advanced earlier (Section 3.2) but it seems pertinent to reproduce the exact words of Bahn and Flenley (1992) on this subject, as climate change is a central point of the present book and will be extensively discussed in the following chapters:

> There could be explanations other than overexploitation of resources. For example, there could have been a major drought. But it seems odd that the forest should survive for at least 37,000 years, including the major climatic fluctuations of the last ice age and the postglacial climatic peak, only to succumb to drought after people arrived on the island. That would be a coincidence just too great to believe. There could have been an invasion of a new group of people, or a disease introduced. But neither would explain all the island's history nor is there independent evidence for either.

The book by Bahn and Flenley (1992) ends with a warning for the whole Earth. They considered that Easter Island may be a microcosmic model for the whole planet, as both are isolated systems with limited resources. Therefore, Easter Island may be viewed as a real experiment on the consequences of overexploitation of natural resources, from which we can extract important lessons for the future of Earth. However, Bahn and Flenley (1992) devised some difficulties for these lessons to be learned. One is human nature and the other is the current attachment to unlimited growth. The first may be illustrated by a paragraph that in some places is erroneously attribute to Diamond (2005) but is really from Bahn and Flenley (1992):

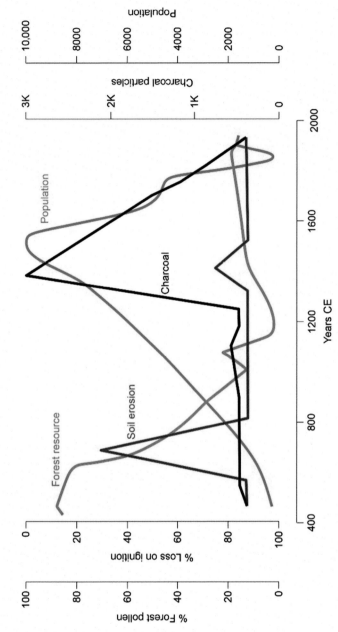

Fig. 4.11 Landscape and demographic inferences for Easter Island based on the Rano Kao pollen record (Flenley et al., 1991), which suggested a series of events starting with forest overexploitation and ending in sudden population decline. *(Redrawn from Bahn, P., Flenley, J., 1992. Easter Island, Earth Island. Tames & Hudson, London.)*

The person who felled the last tree could see that it was the last tree. But he (or she) still felled it. This is what is so worrying. Humankind's covetousness is boundless. Its selfishness appears to be genetically inborn. Selfishness leads to survival. Altruism leads to death. The selfish gene winds. But in a limited ecosystem, selfishness leads to increasing population imbalance, population crash, and ultimately extinction… It seems that human personality is against us. The feller of the last tree knew it would lead ultimately to disaster for subsequent generations but went ahead and swung the axe.

The other drawback is that the present-day global economic system is based on the fallacy of the unlimited growth, which is clearly unsustainable, but nobody is willing to lower its living standards to pave the way toward a more sustainable world. Bahn and Flenley (1992) compared our present "gods" of economic growth and rising standards of living, along with the competition that both involve, with the Rapanui giant statues and all their social connotations. According to these authors, we should throw down our economic moai to regain the sense of proportion that we have lost and to return to a new "religion" centered on the environment. This is another lesson that may be learned from Easter Island, where the shift from the moai cult to the Birdman cult represented a change to more natural and sustainable worship. Finally, Bahn and Flenley (1992) asked themselves:

Will we Easter Islanders have the sense to do the same thing before our skyscrapers came tumbling about our ears? Or is the human personality always the same as that of the person who felled the last tree?

These final words clearly expressed the conviction of Bahn and Flenley (1992) that the Easter Islanders found a solution for surviving—i.e., the shift from the moai cult to the Birdman cult—and did not disappear as a society, as many further interpretations seemed to take for granted. Indeed, Flenley et al. (1991) and Bahn and Flenley (1992) never referred to the disappearance of the Rapanui society but to the end of the megalithic culture. According to these authors:

Yes, there was war and famine, and yes, there was a population crash, but ultimately the Easter Islanders were reconciled with nature and achieved, for a time [here the authors probably refer to the time before European contact], a sustainable stability.

4.5 Some insights on the pioneering works

The work of Flenley and his colleagues was taken as the empirical confirmation of Easter Island's deforestation provoked by the first Polynesian settlers—i.e., the prehistoric Rapanui—which purportedly caused their own cultural collapse. This strongly supported the collapse ideas of Mulloy (1974) explained in Section 2.4.5 and contributed to its establishment as the paradigm for Easter Island's prehistory, which is still followed by many scholars and is the most popular theory among the general public interested in the topic. Flenley's work was, and still is, considered for many as the last word in Easter Island's paleoecology but, as usual in scientific research, the persistence of some knowledge gaps and the further finding of novel empirical evidence has challenged the paradigm and has provided new scenarios for Easter Island's prehistory (Rull et al., 2010). In spite of this, Flenley and his coworkers set the first stone of the paleoecological edifice and influenced all paleoecologists working on Easter Island. The study of this tiny and extremely isolated island in the middle of the Pacific Ocean would never be the same without their wisdom and inspiration. This section is aimed at discussing some unclear features of the pioneering work presented above. The purpose is not to criticize the pioneers but to highlight some aspects that remained unresolved and that have boosted further paleoecological research. Only the literature available by the time of Flenley et al. (1991) will be utilized in this discussion.

4.5.1 Deforestation chronology

Dating and age–depth modeling of the above cores are key aspects, as they were the basis for the definition of the age and temporal trends of island deforestation. The most problematic age–depth model seems to be that of core RRA3 from Rano Raraku. As discussed above, Flenley et al. (1991) developed a single linear model for this sequence after removing some "anomalous" dates (Fig. 4.1) and deduced the age of the onset of deforestation (1200 ^{14}C yr BP) from this model. This chronology is of fundamental importance for the prehistory of Easter Island and merits additional discussion. First, it is very important to note that the interpolated age of 1200 ^{14}C yr BP falls between two contiguous ages that are only 25 cm apart but have almost 6400 ^{14}C yr of difference (Fig. 4.2; Appendix 3.1). This pattern is typical when a sedimentary gap (hiatus) occurs, a situation not contemplated by Flenley et al. (1991). The other part of the record is compatible with an exponential or a locally interpolated age–depth models, as those used by Flenley et al. (1991) in Kao and Aroi, respectively (Figs. 4.3 and 4.5). In this alternative framework, all radiocarbon dates make perfect chronological sense and there is no need for removing any of them (Fig. 4.12). The main drawbacks of this alternative model are the difficulty of dating the deforestation onset and unravelling their temporal trends, as the corresponding sediments would be absent due to the hiatus. Therefore, in Raraku, the

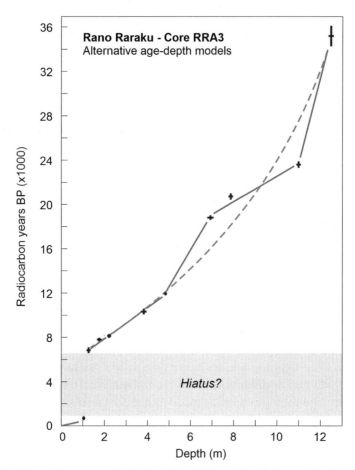

Fig. 4.12 Alternative exponential *(dotted blue line)* and locally interpolated *(solid blue line)* age-depth models for Raraku core RRA3. The horizontal *red* bars represent the length of the sample dated and vertical *red* bars are the dating error. The *green* line represents the accumulation rate of the upper 70 cm of the core after ^{210}Pb analysis. The *gray* band represent a potential depositional gap (hiatus). See Fig. 4.1 for comparison.

deforestation could have occurred at any time between about 7000 and 500 ^{14}C yr BP, at unknown rates. The situation seems clearer at Rano Kao, on which Flenley and coworkers based most of their views on deforestation and cultural collapse (Fig. 4.11).

4.5.2 Paleoclimatic inference and paleoecological implications

As discussed in the former chapter (Section 3.2), the possible influence of climate changes on socioecological developments of Easter Island's prehistory

has been a hot topic for a long time. Flenley et al. (1991) used local paleoecological evidence to demonstrate, for the first time, that Easter Island actually experienced climate changes during Late Pleistocene and Holocene
times. These authors utilized pollen data as paleoclimatic proxies for semi-
quantitative temperature and moisture balance estimation (Section 4.3.2),
which gives rise to two main points for discussion: the reliability of estimates and the usefulness of the resulting paleoclimatic trends for ecological
analysis. Regarding the first point, the indices proposed by Flenley et al.
(1991) to estimate past climatic trends were based on extremely broad taxonomic categories such as tribes (Tubuliflorae), families (grasses, palms), and
even classes (ferns). It should be clarified that this was due to the impossibility of more precise taxonomic identification from pollen found on sediments. However, broad taxonomic categories are in general unsuitable for
climatic reconstruction, as every species has its own climatic requirements,
response lags and migration rates, and responds differently to the same climatic forcing (Wright, 1977; Davis, 1981). In addition, Flenley et al. (1991)
used worldwide biogeographical features of the broad taxonomic categories involved, and the application of these global environmental features
to Easter Island's species is unwarranted. Therefore, the temperature and
moisture indices of these authors should be viewed as very preliminary attempts of paleoclimatic reconstruction. Another important point is that, for
obvious reasons, paleoclimatic trends inferred from plant proxies cannot be
used to analyze vegetation responses to climate change. Flenley et al. (1991)
did not do it but the danger is latent to take paleoclimatic reconstructions
based on vegetation proxies as a matter of fact and use them as standard
paleoclimatic curves to analyze vegetation responses to environmental shifts
(e.g., Hunter-Anderson, 1998). To avoid this circularity, paleoclimatic reconstructions should be based on proxies independent from the organisms
whose response is being analyzed. The most usual of these independent
proxies are physicochemical indicators and remains of organisms of short
life cycles (commonly planktonic or other small aquatic organisms of annual life cycles), which have an almost immediate response to climatic shifts.

4.5.3 The upper altitudinal forest limit

John Flenley was fully aware of the usefulness of oscillations in the elevation of the upper boundary of montane forests (as reconstructed from
pollen analysis) as paleoclimatic indicators (Flenley, 1979a). For example,
in the tropical South American Andes, Van der Hammen (1974) reported
an LGM downward shift of the upper forest line of more than 1000 m, as

a response to a cooling of about 8°C, followed by a progressive upward to its present position during the general Holocene warming. In the tropical Andes, past elevational oscillations in the upper forest limit, or tree line, were reconstructed from fluctuations in pollen from selected trees, after modern-analog studies showing a quantitative relationship between pollen percentages of these trees and elevation. As already mentioned, modern-analog studies are impracticable on Easter Island (Section 1.4.4) and Flenley et al. (1991), who were already aware of this, based their inferences on the ecological and biogeographical information existing for the most probable parent taxa of the sedimentary pollen encountered (Appendix 4.1). Using this information, these authors suggested that the upper forest belt could have been dominated by *Coprosma* and an unknown Compositae-Tubuliflorae species and have been oscillated up and down around the elevation of Rano Aroi, depending on temperature fluctuations (Fig. 4.10). Again, due to the total destruction of the original vegetation of the island, this point is hard to be demonstrated and should be taken with great care.

4.5.4 Human and climatic impact

After Flenley et al. (1991), the role of human activities in Easter Island's deforestation seemed unquestionable and was based on several types of evidence. These authors emphasized the chronological coincidence of deforestation and human presence and made special emphasis on the total island's deforestation and the abandonment of the megalithic moai industry, which was dated back to 1680 CE at that time. Paleoecological records also showed an abrupt increase of soil erosion, likely due to the removal of the forest cover, and the onset of fires, probably due to anthropogenic burning of forests. On the basis of this evidence and the observation that Easter Island's forests had survived all the previous Late Pleistocene and Holocene climatic changes, the potential influence of climate on forest clearing was dismissed by these authors. However, they did not consider the possibility of climate changes during the last millennia, whose sedimentological signal would have been obscured or removed by the overwhelming dominance of anthropogenic proxies (and the potential occurrence of a >6000-year sedimentary gap; see Section 4.5.1), even in the geochemical record. It is possible that Flenley et al. (1991) considered recent minor climatic shifts such as the MCA or the LIA

(Section 3.2) negligible in an island where a major worldwide cooling such the LGM had been of less than 2°C. However, this is only a speculation. The truth is that these authors did not discuss the possibility of climatic changes after human settlement.

4.5.5 The socioecological catastrophe

The socioecological synthesis by Bahn and Flenley (1992) relied on the still untested assumption that humans were only responsible for deforestation and cultural collapse, and that climate change did not play any role in these events. In addition, the chronological coincidence between total deforestation (500 ^{14}C yr BP, approximately 1450 CE) and the end of the megalithic culture (1680 CE) is questionable, although Flenley et al. (1991) considered these dates to be "not too far" (Section 4.3.3). Therefore, erecting a socioecological collapse theory as illustrated in Fig. 4.8 on the basis of these uncertain premises seems, at least, premature. The Easter Island collapse theory, however, was very welcome by a vast majority of the scientific and popular audiences likely because of its worldwide environmental and socioeconomic implications, rather than for its empirical foundations. In other words, the Easter Island's collapse theory, whether true or not, arrived at the right time and was deliberately utilized to emphasize the need for a reconsideration of the current overexploitation of natural resources, at a global level, under a landscape of unlimited economic growth. As laudable as this initiative may be, socioeconomical and political success (or convenience) is not the right way of testing scientific hypotheses, whose acceptance or rejection should be based on empirical data. Therefore, the points highlighted in this section, and others, should be revisited for a sound appraisal of what really happened on Easter Island.

Appendix 4.1

Pollen types, probable parent plants and ecological/biogeographical notes, according to Flenley et al. (1991). Some taxonomical, ecological, and/or biogeographical information may differ from those presented in Chapter 1 (Section 1.4), which is more updated than the one used by Flenley et al. (1991). A, Aquatics; H, Herbs; Pt, Pteridophytes; S, Shrubs; T, Trees.

Pollen types	Habit	Family	Probable parent plants	Ecology and biogeography
Acalypha *Canavalia* *Capparis* comp.	TS H TS	Euphorbiaceae Leguminosae Capparidaceae	*Acalypha* sp. *Canavalia* sp. *Capparis* sp.	Widespread shrubs in SE Asia and the Pacific A widespread herb, especially coastal Not recorded in Easter Island, but widespread in the Pacific as a coastal shrub
Caryophyllaceae	H	Caryophyllaceae	*Cerastium glomeratum,* *Dianthus caryophyllus,* *Paronychia* sp., *Polycarpon tetraphyllum*	An introduced species. *Paronychia* occurs on Juan Fernández Island
Casuarina	TS	Casuarinaceae	*Casuarina equisetifolia*	Introduced tree, widespread in the Pacific. The pollen is capable of long-distance dispersal from Australia
Compositae- Tubuliflorae	TS	Compositae	*Campylotheca* sp., *Chrysogonum* sp.	Probably Heliantheae. There are no native representatives of this tribe in the present flora, but these two genera occur on other islands in SE Polynesia
	H		*Ageratum conyzoides,* *Bidens Pilosa, Centaurea cyanus, Cirsium vulgare,* *Conyza bonariensis,* *Cotula australis, Cynara scolymus, Erigeron linifolius, Galinsoga parviflora*	Introduced herbaceous species which may well contribute to the recent part of the record

		Family	Species	Notes
Compositae-liguliflorae	H	Compositae	*Hypochoeris radiata, Lactuca sativa, Sonchus asper, Sonchus oleraceus, Taraxacum officinale*	Introduced species now widespread on the island
Coprosma	TS	Rubiaceae	*Coprosma* sp.	Extinct on the island, but occurs as a shrub or small tree in the upper forests of many high Polynesian islands, including Rapa. Also on Juan Fernández
Cyperaceae-*Sirpus* comp.	A	Cyperaceae	*Sirpus californicus*	The dominant of all three crater swamps. Also in South America. The genus is widespread in the Pacific
Cyperaceae-*Cyperus* comp.	A	Cyperaceae	*Cyperus cyperoides, C. polystachius, C. vegetus, Kyllinga brevifolia*	*Cyperus* spp. Occur marginally in the crater swamps. *Kyllinga brevifolia* is widespread over the island
Ephedra	TS	Ephedraceae	*Ephedra* spp.	South America is the likely source, by long-distance dispersal
Euphorbia comp. *serpens* sim.	H	Euphorbiaceae	*Ephorbia serpens, E. hirta, E. peplus*	The first two are likely native. *E. serpens* is especially common on overgrazed land
Filices-monolete psilate	Pt	Filices	*Filices*	Almost all monolete fern spores could be included here when they have lost the perine
Filices-monolete verrucate/areolate	Pt	Aspidiaceae	*Dryopteris espinosa*	Endemic; grows among rocks
			D. gongyloides	Pantropical; grows in the crater swamps
			D. parasitica	Widespread; grows among rocks
		Aspleniaceae	*Asplenium adiantoides*	Widespread; grows in swamps
		Davalliaceae	*Davallia solida*	Genus widespread; grows on rocks in shade
		Polypodiaceae	*Polypodium scolopendria*	Widespread; grows among rocks
		Lomariopsidaceae	*Elaphoglossum skottsbergii*	Polynesia; grows among rocks

Continued

Pollen types	Habit	Family	Probable parent plants	Ecology and biogeography
Filices–monolete scabrate	Pt	Blechnaceae	*Doodia paschalis*	Endemic; grows among rocks
		Vittariaceae	*Vittaria elongata*	Widespread; grows on peat, sometimes in swamps
		Aspleniaceae	*Asplenium obtusatum*	Widespread; coastal
		Psilotaceae	*Psilotum nudum*	Widespread; recorded near Rano Aroi
Filices–trilete psilate	Pt	Dennstaedtiaceae	*Microlepia strigosa*	Widely distributed in Polynesia. Common on the island. May include other trilete spores which lost the perine
Filices–trilete echinate	Pt	Cyatheaceae?	*Cyathea* sp.?	Widespread. Tree ferns in forest on other Polynesian islands
Filices–trilete scabrate	Pt	Cyatheaceae?	*Cyathea* sp.?	Widespread. Tree ferns in forest on other Polynesian islands
Filices–trilete verrucate	Pt	Cyatheaceae?	*Cyathea* sp.?	Widespread. Tree ferns in forest on other Polynesian islands
Gramineae	H	Gramineae	46 species	Perennial grasses which form the dominant vegetation over most of the island today
Lycopodium–reticulate	Pt	Lycopodiaceae	*Lycopodium* sp.	Genus common in the forests on the high Polynesian islands
Lycopodium–foveolate	Pt	Lycopodiaceae	*Lycopodium* sp.	Genus common in the forests on the high Polynesian islands
Melia comp.	TS	Meliaceae	*Melia azedarach*	Widespread in the tropics. Introduced and now common in the island

		Family	Species	Notes
Macaranga comp.	TS	Euphorbiaceae	Macaranga sp.	Not recorded on the island, but common in SE Asia. The pollen could arrive by long-distance dispersal
Myrtaceae–Psidium sim.	TS	Myrtaceae	Psidium guajava	Widespread in the tropics. Introduced and now common on the island
Myrtaceae undiff.	TS	Myrtaceae	Metrosideros sp.?	Widespread in the Pacific. Absent from Easter Island, but could have occurred in the past
Ophioglossum	Pt	Ophioglossaceae	Ophioglossum coriaceum	Widespread; grows in short grass
Palmae	TS	Palmae	Extinct species related to Jubaea chilensis	J. chilensis is a Chilean endemic, valued for its edible fruit and sugary fermentable sap
Plantago id. lanceolata comp.	H	Plantaginaceae	Plantago lanceolata	Introduced weed
Polygonum id. acuminatum comp.	A	Polygonaceae	Polygonum acuminatum	Widespread in South America. Marginal to all three crater swamps. Used as a medicinal plant
Potamogeton	A	Potamogetonaceae	Potamogeton sp.	Not recorded on the island
Pteris comp.	Pt	Pteridaceae	Pteris sp.	
Rubiaceae	TS	Rubiaceae	Rubiaceae spp.	Likely a woody Rubiaceae arrived by long-distance dispersal from some Polynesian island (a single grain found)
Rumex sim.	H	Polygonaceae	Rumex obtusifolius	Introduced
Sapindus	TS	Sapindaceae	Sapindus saponaria	Introduced
Solanum comp.	H	Solanaceae	Solanum forsteri	Introduced. Reported to have been cultivated in historic times
	S		Lycium carolinianum	Native
Sophora	TS	Leguminosae	Sophora toromiro	Endemic, extinct on the island since the 1960s but survives in Goteborg Botanic Garden from seeds collected by Heterdahl. Attempts at reintroduction have recently been made

Continued

Pollen types	Habit	Family	Probable parent plants	Ecology and biogeography
Trema comp.	TS	Ulmaceae	*Trema* sp.	Widespread in SE Asia and the Pacific. Not clear whether the pollen indicates long-distance dispersal or occurrence of the taxon in the past. In primary and secondary forest
Trumfetta comp.	TS	Tiliaceae	*Triumfetta semitriloba*	Rare on Easter Island but widespread in the Pacific. Used to make rope
Typha angustifolia sim.	A	Typhaceae	*Typha angustifolia*	Not recorded on Easter Island, but could have occurred in the past. Occurs in the Pacific (Fiji and New Zealand) and in South America
Umbelliferae comp. *Apium* sim.	H	Umbelliferae	*Apium amni*	Introduced
			A. prostratum	Native
			Foeniculum vulgare	Introduced
Urticaceae/ Moraceae	TS	Moraceae	*Broussonetia papirifera*	Introduced. Widespread in the Pacific and used to make tapa (bark cloth). It survives on Easter Island as an occasional shrub
			Ficus carica	Introduced
			Morus sp.	Introduced

CHAPTER 5

The transitional phase: Paleoecological impasse

Contents

5.1	Rano Kao and the dating problem	169
5.2	Rano Raraku	172
	5.2.1 Paleolimnology and the Amerindian influence	172
	5.2.2 Soil and vegetation degradation	175
5.3	Rano Aroi: More dating problems	178
5.4	Other studies	178
5.5	Paleoecological impasse	181

After the apparent consolidation of Mulloy's (1974) collapse theory by Flenley et al. (1991) and Bahn and Flenley (1992), paleoecological research on Easter Island experienced an impasse of roughly a decade, which is called here the transitional phase (Fig. 3.7). This transition was characterized by sporadic, nonsystematic coring of the ranos and the publication of a few papers that did not introduce significant modifications to the collapse paradigm. During this phase, Flenley and his team focused on core KAO2 (Fig. 3.6; Table 3.1), from Rano Kao, which had been retrieved in the 1983 campaign but remained unstudied. Other few short cores possibly encompassing the last centuries/millennia from Rano Aroi and Rano Raraku were also retrieved by other researchers (Peteet et al., 2003; Mann et al., 2003). This chapter summarizes the few paleoecological studies carried out during this intermediate phase, sorted by sites and in chronological order of publication.

5.1 Rano Kao and the dating problem

Flenley (1996, 1998) considered that Rano Kao was particularly interesting for several reasons. First, their inner crater walls may have created a favorable microclimate for an unusual type of forest. Second, the pioneering studies (Chapter 4) suggested that the Kao catchment might have been deforested early in the colonization of the island. Third, the same earlier studies demonstrated the unsuitability of the younger sections of the other ranos for providing a reliable account of the last centuries due to either physical sediment disturbance (Rano Aroi) or contamination by older carbon (Rano Raraku). For these reasons, studies previously developed on the marginal KAO1 core were extended to core KAO2, which had been taken in a more central position (Fig. 3.6). This core was 20.85 m depth and contained three well-differentiated parts. The uppermost 3.5 m corresponded to the floating mat and were followed by a water gap of 7 m, and a lowermost sequence of bottom sediments of 10.35 m (Flenley, 1996). Radiocarbon dating of core KAO2 exhibited some anomalies, notably two age inversions. The upper inversion occurred between the bottom of the floating mat and the upper part of the bottom sediments (Fig. 5.1), and was attributed to the drifting of

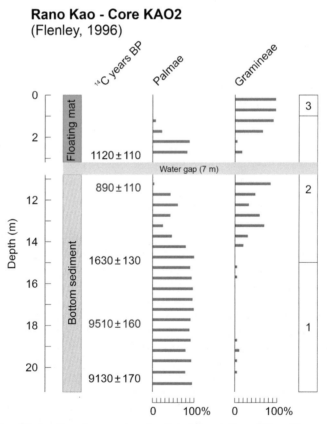

Fig. 5.1 Simplified pollen diagram showing the deforestation of Rano Kao, as recorded in core KAO2 (Flenley, 1996). Radiocarbon ages considered to be anomalous are highlighted in *red*.

younger material beneath the floating mat. The lower inversion occurred in the bottom sediments and was explained in terms of contamination of the oldest sample (9510 ± 160 ^{14}C yr BP) with older carbon probably from the inwash of soils from the steep crater walls as a consequence of initial human disturbance. Therefore, this date was rejected (Flenley, 1996).

The pollen diagram of KAO2 (Fig. 5.1) was initially subdivided into three zones representing three different vegetation types. Zone 1 represented a catchment totally dominated by palm forests but its chronological extent remained uncertain. These forests started to decline in Zone 2 and were progressively replaced by grass meadows. Therefore, the date of 1630 ± 130 ^{14}C yr BP (calibrated to approximately 150–680 CE) may indicate the arrival of Polynesian people in the island. This date was considerably earlier than 1200 ^{14}C yr BP, the date obtained in KAO1 for the start of forest clearance (Section 4.3.1). According to Flenley (1996), this was not surprising, since KAO1 was close to the lakeshore and recorded mostly local phenomena, whereas KAO2 was more central and provided a more overall picture of what happened in the catchment. Zone 3, starting at about 400 ^{14}C yr BP (interpolated age), represented the present treeless landscape of Rano Kao crater catchment and of the whole island. Again, this date was different from that obtained at KAO1 for total deforestation (550 ^{14}C yr BP; Section 4.3.1), which was attributed to the same reason mentioned above regarding the location of coring sites. Shortly after the publication of the first analyses of core KAO2, Flenley (1998) explicitly considered the possibility of some climatic shifts, probably LIA droughts, as previously suggested by McCall (1993), to have been involved in the cultural collapse. However, the main handicap was the lack of local and independent evidence (Section 4.5.2) for the occurrence of such droughts on Easter Island.

Using the same core, Butler and Flenley (2001) and Butler et al. (2004) increased the number of dates and the variety of dated materials to refine the age-depth model. Before analyzing these data in more depth, it is important to note that Butler and Flenley (2001) analyzed separately the floating mat and the bottom sediments and found a general correlation between the lower and the upper parts of these sequences, something that was already visible in previous analyses (Fig. 5.1). Indeed, the upper part of the bottom sediments looks like a repetition of the floating mat sequence, which suggests that both accumulated at the same time (Section 3.3.2) and likely records the same deforestation process, rather than two consecutive forest clearing events. This was not discussed by Butler and Flenley (2001), who avoided any chronological considerations until more radiocarbon dates were available. In their own words: "In the absence of further radiocarbon dating, it is impossible to draw firm conclusions about the ecological history of the island…." A latest attempt to obtain a reliable age-depth model for core KAO2 was made using 18 radiocarbon dates on bulk sediments, and several sediment fractions such as large plant fragments, tiny plant fragments, and pollen concentrates (Butler et al., 2004). In this case, radiocarbon dates were calibrated to calendar years using INTCAL98 (Stuiver et al., 1998).

The results showed a scattered and rather inconsistent pattern that prevented to establish a reliable age-depth model for this core (Fig. 5.2). Butler et al. (2004) discussed several potential contamination sources to explain the presence of "too old" carbon in many samples. Inwash from the inner crater walls, as previously suggested for KAO1 core (Flenley et al., 1991), was dismissed as a major contamination source, as the coring site was 300 m away from the lakeshore. Other possible sources considered were fine particles derived from the burning of aged trees outside the crater, depleted CO_2 emitted as gas from the caldera floor and incorporated by photosynthesis, or erosion of marginal lake sediments due to the periodic occurrence of low lake levels during climatic droughts. The authors concluded that the suitability of floating mats as paleoecological archives was doubtful, due to the possibility of these mats to be inverted or "flipped over" in the past, and that bottom lake sediments were better suited to correlate the Rano Kao palynological records with those from Rano Raraku (Butler et al., 2004). These authors recommended caution in the interpretation of the KAO2 palynological record available at that time and the development of additional coring and new chronologies after careful selection of the materials to be dated. In summary, the new studies developed on Rano Kao during the transitional phase did not contribute to clarifying the spatial and temporal patterns of deforestation. In spite of this, Butler and Flenley (2001) claimed that these new results confirmed the general results obtained in the pioneering phase, in particular, that (i) the island was forested for many thousands of years before human impact, (ii) climate change had only a minor impact on the vegetation, and (iii) human impact was the major cause of the decline of forest.

5.2 Rano Raraku

5.2.1 Paleolimnology and the Amerindian influence

A 3.40-m sediment core (SW) was taken in 1990 by Dumont et al. (1998) in Rano Raraku and analyzed for some physicochemical (e.g., magnetic susceptibility) and biological proxies (e.g., cladocera, ostracods, diatoms,

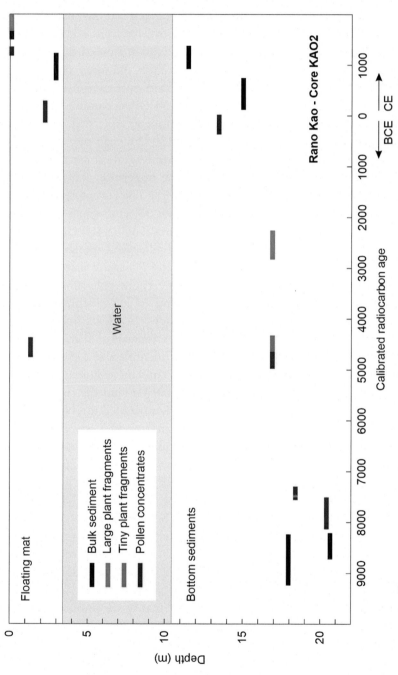

Fig. 5.2 Calibrated ^{14}C ages versus depth for all dated fractions of the Rano Kao core KAO2 (Butler et al., 2004). Bar length represents 2σ interval. *(Redrawn and modified from Butler, K., Prior, C.A., Flenley, J.R., 2004. Anomalous radiocarbon dates from Easter Island. Radiocarbon 46, 395–420.)*

chrysophytes, and fossil pigments). Radiocarbon dating of two samples between about 1.6 and 2.2-m depth produced "much too old" dates (c.16–24 ^{14}C kyr BP) (Dumont et al. 1998). These dates were rejected using the same arguments of Flenley et al. (1991), who attributed the old ages of some of their samples to massive inwash from the inner crater walls (Section 4.3.1). A sample situated at 1.30–1.35 m gave an age of 588 ± 60 ^{14}C yr BP, which was calibrated to 1300–1450 CE, according to Stuiver and Reimer (1986). After ^{210}Pb dating, Dumont et al. (1998) concluded that sediments at 100 cm depth were at least 150–200 years old. Gross pollen analysis using only palm vs. nonpalm (mainly grasses) categories was used to correlate the sediment core with the uppermost 220 cm of the Raraku core (RRA3) from Flenley et al. (1991). The abrupt replacement of palm by nonpalm pollen occurred at about 130 cm, which coincided with the formerly calibrated radiocarbon age of 1300–1450 CE. In their biological analyses, Dumont et al. (1998) found several sponge (*Meyenia* sp.), diatom (*Navicula goeppertiana* var. *monita* and *Pinnularia latevittata*), and chrysophyte species of Neotropical origin appearing synchronously above 135 cm depth in the sediments. Another wave of introductions including subantarctic and Australian species of cladocera (*Alona weinecki*), ostracods (*Sarscypridopsis* cf. *elisabethae*), diatoms (*Achnanthes* cf. *abundans* and *Nitschia* cf. *vidovichii*), and chrysophytes was identified at 115 cm depth. Using the above chronology and assuming a constant sedimentation rate, the Neotropical event was dated to 1300–1450 CE and the subantarctic event to 1527–1685 CE. Dumont et al. (1998) attributed the Neotropical immigration wave to the arrival of Amerindians to Easter Island long before the European contact. Regarding the subantarctic wave, these authors considered that the dates of 1527–1685 CE were approximate, as based on linear interpolation, and suggested that the subantarctic fauna might have been introduced by some of the first European visitors, probably James Cook and his crew, who arrived in 1774 (see Section 2.1).

Another point raised by Dumont et al. (1998) was the presence of *Scirpus californicus* (totora) on Easter Island. As we saw in Chapter 4, Flenley et al. (1991) recorded the presence of this species since about 38,000 ^{14}C yr (e.g., Fig. 4.6) but Dumont et al. (1998) argued that this was based on two uncertain premises, namely the acceptance of questionable radiocarbon dates and the morphological differentiation of *Scirpus* pollen from that of other autochthonous Cyperaceae species (see Table 1.5), for which Flenley et al. (1991) did not provide the corresponding identification criteria. Based on macrofossil evidence, which was considered by Dumont et al. (1998) more reliable than sedge pollen identification, these authors proposed that there was no proof of the presence of totora on Easter Island before the 14th century. The chronological coincidence of this first occurrence of totora with the Neotropical wave of plankton introduction mentioned above (1300–1450 CE), suggested the authors that the totora may have also been introduced by Amerindians, who used this species in the construction of their ships. The whole picture would be consistent with the newcomer hypothesis that, as explained in Section 2.3.3, has recently been supported by DNA analysis of modern Rapanui people living on the island (Thorsby, 2016). Dumont et al. (1998) also suggested that Amerindians—the Incas, according to these authors—may have contributed to the cultural collapse of the island by halting the tradition of moai carving, which is just the opposite of the proposed by Heyerdahl (Section 2.3.2).

5.2.2 Soil and vegetation degradation

In an attempt to document the timing of earliest vegetation clearance by humans and its impact on soils, Mann et al. (2003) combined paleoecological and edaphological methods. These authors developed an island-wide edaphic survey (Fig. 3.12), including soil stratigraphy and radiocarbon dating, and compared the results with a new sedimentary record from Rano Raraku. Primeval soils—i.e., soils that supported the former Easter Island's forests—were recognized by the characteristic presence of palm root casts up to 2 m depth and 1 cm thick. Remarkably, these authors did not find primeval soils of this type in the southern slopes of Terevaka, above 300 m elevation, which would support the palynological interpretation of Flenley et al. (1991) about these forests as typically lowland forests (Section 4.3.1). Primeval soils were often buried but could be recognized in soil pits and naturally exposed sections in foot-slope areas and along erosional gullies. In these sections, a prominent unconformity existed between primeval soils, which were truncated by erosion, and the overlying slope wash deposits (Fig. 6.7). The age of soil truncation was considered to be an estimate of the timing of forest clearance and the end of primeval soil development. Mann et al. (2003) dated charcoal samples from the overlying deposits and observed that none of these dates predated 1200 CE, which suggested that primeval soils had not been truncated until this date. In addition, almost all charcoal dates were situated between 1200 and 1650 CE, which suggested that a major event of island-wide primeval soil truncation occurred during these 450 years, roughly coinciding with the phase of megalithic

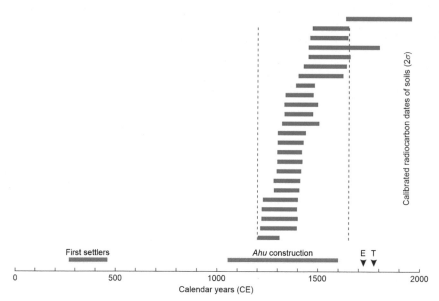

Fig. 5.3 Calibrated radiocarbon ages of charcoal from soils overlying the primeval forest soils *(blue bars)* and main prehistorical events *(orange bars and red arrows)*. E, European contact; T, *Moai* toppling. *(Redrawn and modified from Mann, D., Chase, J., Edwards, J., Beck, W., Reanier, R., Mass, M., 2003. Prehistoric destruction of the primeval soils and vegetation of Rapa Nui (Isla de Pascua, Easter Island), In: Loret, J., Tanacredi, J.T. (Eds.), Easter Island. Scientific Exploration into the World's Environmental Problems in Microcosm. Kluwer Academic/Plenum, New York, pp. 133–153.)*

statue construction (Fig. 5.3). The presence, in the overlying deposits, of abundant charcoal and obsidian flakes (mata'a) was interpreted in terms of massive deforestation across the island using slash-and-burn techniques (Section 2.4.5).

The parallel study of a gravity core from Rano Raraku, not named by Mann et al. (2003), provided some supporting information. Indeed, a sudden increase in sedimentary charcoal occurred at 1070–1280 CE—a date obtained by radiocarbon dating of totora seeds and calibration after Stuiver et al. (1998)—suggested massive forest burning (Fig. 5.4). At the same time, magnetic susceptibility abruptly increased and sedimentary organic matter sharply declined, which was consistent with enhanced soil erosion and clastic influx to the lake, probably due to vegetation removal and human activities. Mann et al. (2003) concluded that widespread forest clearance began on Easter Island only after 1200 CE and asked why this deforestation was delayed for centuries after initial settlement, which they situated by 300 CE. The authors speculated that before 1200 CE the island was occupied,

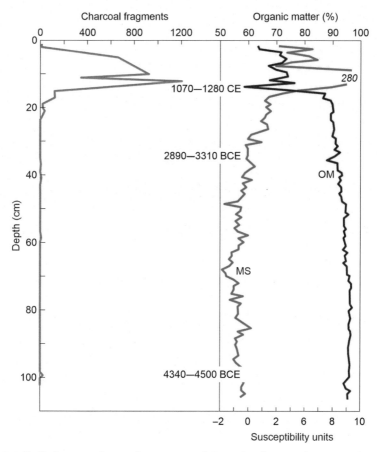

Fig. 5.4 Preliminary analyses of an unnamed Rano Raraku core (Mann et al., 2003) showing the charcoal content *(gray line)*, the organic matter content *(red line)*, and the magnetic susceptibility *(blue line)*. Chronology was based on calibrated radiocarbon dates obtained in *Scircpus californicus* (totora) seeds. *(Redrawn from Mann, D., Chase, J., Edwards, J., Beck, W., Reanier, R., Mass, M., 2003. Prehistoric destruction of the primeval soils and vegetation of Rapa Nui (Isla de Pascua, Easter Island), In: Loret, J., Tanacredi, J.T. (Eds.), Easter Island. Scientific Exploration into the World's Environmental Problems in Microcosm. Kluwer Academic/Plenum, New York, pp. 133–153.)*

perhaps transiently, by Polynesian hunter-gatherers thriving on its rocky shores. Permanent establishment, dryland farming, and population growth might have initiated by 1200 CE, leading to the ecological transformation of the island (Section 2.3.2). Using similar edaphic evidence, Mieth et al. (2002) and Mieth and Bork (2003, 2005) dated the beginning of forest clearance in the SW Poike peninsula (Fig. 3.12) by 1280 CE.

5.3 Rano Aroi: More dating problems

The Aroi swamp was also cored during the so-called transitional phase but dating problems similar to those of Rano Kao prevented reliable paleoecological reconstruction. An 8-m core (RA2) was obtained in the northernmost part of Rano Aroi using a Livingstone corer (Fig. 3.6; Table 3.1). The whole core was analyzed for Diffusal Spectral Reflectance (DSR) but only the 200–300 cm interval (Drive 3) was dated and used for paleoecological (palynological) analysis (Peteet et al., 2003). These authors assumed that the interval analyzed corresponded to the Late Glacial-Holocene transition but this view seems hard to sustain with the radiocarbon dates provided (Fig. 5.5). Pollen and spores exhibited major fluctuations along the section studied but the absence of a reliable chronology prevented to follow the timeline of vegetation dynamics. For this reason, only general considerations and comparisons with the previous work of Flenley et al. (1991) were made, and no time-ordered paleoecological reconstruction was possible.

5.4 Other studies

Phytolith analysis is widely used in archaeology as it provides information on the human use of plants, as well as on several paleoecological features (Piperno, 2006). On Easter Island, the pioneer of phytolith studies was Cummings (1998), who worked on selected samples from three archaeological sites: Te Niu, Ahu Heiki'i, and La Pérouse (Fig. 3.12). This work was aimed mainly at identifying garden areas and the crops that were grown in them. These pioneering analyses demonstrated that the preservation of pollen and starch granules of economically important plants for the Rapanui—e.g., *Colocasia, Cordyline, Curcuma, Ipomoea, Broussonetia, Musa,* and *Triumfetta* (Section 1.4.3)—was good enough to reconstruct their agricultural practices. In addition, phytolith analysis provided insights on the surrounding vegetation. Palmae phytoliths resulted to be dominant in all samples analyzed, which indicated the prevalence of palms in the local vegetation at some time. Samples were not dated but the archaeological context suggested that some of them could correspond to the periods 1400–1450 and 1300–1700 CE (Cummings, 1998). However, the author warned that palms produce huge amounts of phytoliths in virtually all parts of the plant and that these phytoliths remain on soils even after decomposition of organic plant remains. Therefore, palm phytoliths may be abundant in soils even after the clearing of a palm forest and its replacement by secondary vegetation or crops.

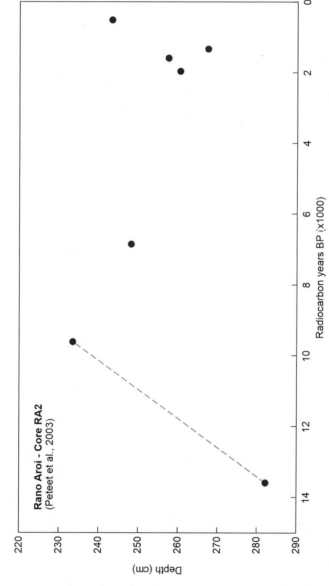

Fig. 5.5 Radiocarbon ages obtained in Drive 3 (200–300 cm) of Rano Aroi core RA2 (Peteet et al., 2003). The dotted *blue* line is the age-depth model chosen by Peteet et al. (2003) by comparison with the Flenley et al. (1991) stratigraphy.

Table 5.1 Woody plant taxa identified from charcoal fragments found in several archaeological sites from three localities (La Pérouse, Akahanga, Orongo) with ages ranging from 610 ± 80 to 220 ± 70 ^{14}C yr BP (Orliac, 2000).

Family	Taxa	Native to EI	Polynesian introduction	Presently absent in EI	Present distribution
Caesalpiniaceae	Caesalpinia cf. major	X			Widespread
Malvaceae	Thespesia populnea		X		Tropics and subtropics
	Broussonetia papirifera		X		Pacific
Sapindaceae	Sapindus saponaria		X		America and Pacific islands
Tiliaceae	Triumfetta semitriloba	X			America
Fabaceae	Sophora toromiro	X			Endemic to Easter Island
Arecaceae	cf. Paschalococos disperta	X			Extinct
Rubiaceae	Coprosma			X	Hawai'i, New Zealand
	Psydrax cf. odorata			X	Pacific (widespread)
Elaeocarpaceae	Elaeocarpus rarotongensis			X	Austral and Cook Islands
Myrsinaceae	Myrsine			X	Eastern Polynesia
Myrtaceae	Syzygium cf. malaccense			X	Pacific (widespread)
Pittosporaceae	Pittosporum			X	Austral Islands
Rhamnaceae	Alphitonia cf. zizyphoïdes			X	Vanuatu, Society, and Marquesas Islands
Verbenaceae	Prenna cf. serratifolia			X	South Pacific
Rubiaceae?	cf. Psychotria			X	East Polynesia
Salicaceae?	cf. Xylosma			X	Pantropical

Unidentified taxa have been excluded. EI, Easter Island.

Other remains of paleoecological interest present in archaeological sites were charred pieces of wood that were identified by anatomical analysis. As mentioned in Section 1.4.3, Orliac (2000) and Orliac and Orliac (1998) performed this type of analysis in three archaeological sites (Fig. 3.12) and identified a number of woody taxa corresponding to the 14th–17th centuries (Table 5.1). Only four of these taxa had been identified at the genus level in preanthropogenic pollen records (*Triumfetta*, *Sophora*, *Coprosma*, and *Paschalococos*), whereas other three were considered to have been introduced by Polynesians (*Thespesia*, *Broussonetia*, and *Sapindus*). However, 10 taxa were of unknown origin and are no longer present on the island. It has been speculated whether these taxa were elements of the ancient forests undetected by pollen analysis because of pollen production and dispersal constraints. This seems unlikely, as no single pollen grain has been found in any of the numerous Easter Island records obtained to date, including the most recent ones (Chapter 6). Another possibility is that these taxa were cultivated by Polynesians for a while and then abandoned. It has also been proposed that the charred woods were actually the remains of drifting wood used by Polynesians for fire.

5.5 Paleoecological impasse

In summary, the paleoecological impasse occurred after the pioneering studies of Flenley and coworkers was characterized by the finding of dating problems (notably age inversions) in Rano Kao and Rano Aroi that, together with the previously detected chronological uncertainties detected in Rano Raraku (Section 4.5.1), complicated the study of island's deforestation and its potential relationship with prehistoric cultural changes. The study by Mann et al. (2003) shed some new light on forest clearing chronology, although the evidence for deforestation was indirect (i.e., soil-charcoal dating and sedimentary shifts in Rano Raraku) and palynological studies on vegetation change were lacking. In spite of this, the unnamed core obtained by these authors seemed promising for further and more complete paleoecological studies. After this impasse, it was difficult to escape the idea that prehistoric paleoecological trends on Easter Island, especially deforestation, were more difficult to elucidate than previously thought. As emphasized by Butler et al. (2004), new coring campaigns and new analyses on the already existing cores were needed for a better appraisal of paleoecological shifts. This occurred during the revival phase, as is explained in the following chapter.

CHAPTER 6

The revival: An opportunity for climate change

Contents

6.1 Coring intensification and reanalysis		183
6.1.1 Rano Raraku		184
6.1.2 Rano Kao		188
6.1.3 Soil investigations		193
6.2 Publication resurgence		195
6.2.1 Polynesian crops		196
6.2.2 Continuous deforestation records		207

The revival phase can be subdivided into two well-differentiated parts. The first part extended between 2005 and 2010 and was characterized by a significant increase of coring activities and the publication of some reanalysis of previously obtained cores (Fig. 3.7). The second part (2012–2019) was marked by the publication of the analyses developed on the cores obtained between 2005 and 2009, which represents an average delay of several years (Section 3.4). The emphasis of this chapter will be on the last millennia, to capture the prehistoric events on which the book is focused.

6.1 Coring intensification and reanalysis

During the first part of the revival phase, a total of 20 cores were retrieved, of which almost half (9) were from Rano Raraku, seven were from Rano Kao, and four were obtained in Rano Aroi (Table 3.1). Some preliminary results of these cores (Azizi and Flenley, 2008; Gossen, 2007; Sáez et al., 2009) along with some reanalyses of older cores (Mann et al., 2008; Butler and Flenley, 2010) were published during the same time interval. All these papers corresponded to Rano Raraku and Rano Kao. During this time interval, soil research also provided interesting paleoecological information (Horrocks and Wozniak, 2008; Mieth and Bork, 2010).

Paleoecological Research on Easter Island
https://doi.org/10.1016/B978-0-12-822727-5.00006-4

6.1.1 Rano Raraku

A new core from Rano Raraku was retrieved by Flenley's team in 2005. This 13.4-m Livingstone core (RRA5) was close to the center of the lake (Fig. 3.6) and only the deepest part (7–13.4 m) was dated and analyzed for pollen and spores. The section studied encompassed approximately 18,000 years (28,000–10,000 ^{14}C yr BP) and included the LGM, which was the main target of the study (Azizi and Flenley, 2008). The authors concluded that the coolest and driest conditions occurred around 17,000 ^{14}C yr BP and that temperatures were c.2°C lower than at present, which supported previous interpretations on older cores (Flenley et al., 1991; Section 4.3.2). These results, however, do not impact on the main target of the book, which is the prehistory.

The preliminary analyses of the unidentified Raraku core retrieved by Mann et al. (2003) in 1998 (Fig. 5.4; Section 5.2.2) were completed with palynological analyses to refine the deforestation chronology, as well as to test the potential impact of rats and climatic droughts on forest clearance (Mann et al., 2008). The formerly unnamed core was identified as "core #1" and another core taken only 3 m apart was named as "core #2" (Fig. 3.6; Table 3.1). Both cores were retrieved using a gravity corer and were about 1 m deep. Radiocarbon dating performed mainly on totora seeds revealed the occurrence of a sedimentary gap between 14 and 16 cm corresponding to the interval 3880±40 to 830±40 ^{14}C yr BP (4090–4410 cal yr BP to 1180–1290 CE) in core #1, and a similar hiatus between 18 and 24 cm corresponding to the interval 4320±40 to 560±40 ^{14}C yr BP (4640–4970 cal yr BP to 1320–1450 CE) in core #2 (Fig. 6.1). Age calibration was carried out using, for the first time on Easter Island, a calibration database (ShCal04) specific for the Southern Hemisphere (McCormac et al., 2004). The resulting age-depth model was very similar to the alternative interpretation proposed in Section 4.5.1 for the core RRA3 of Flenley et al. (1991), although the hiatus was smaller in cores #1 and # 2 of Mann et al. (2008). These authors emphasized this fact and the occurrence of another hiatus in the SW core of Dumont et al. (1998), somewhere between 130 cm (590±60 ^{14}C yr BP) and 165 cm (16,090±70 ^{14}C yr BP). Mann et al. (2008) suggested that a conspicuous depositional hiatus was present in the Raraku sediments probably due to a climatic drought (or a series of droughts) that caused the drying out of the lake, thus interrupting sedimentation for about 3000 years, until 830 ^{14}C yr BP (1180–1290 CE), when the lake replenished again. The drought was attributed to a latitudinal shift in the subtropical storm track (Fig. 1.5), which controls the intensity and frequency of cyclonic

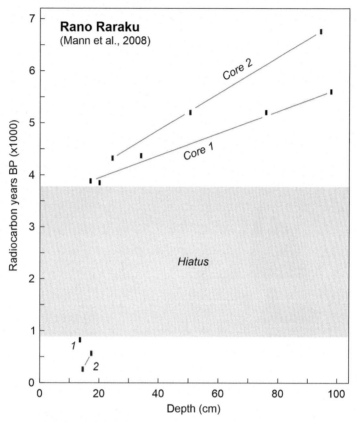

Fig. 6.1 Age-depth plot for cores #1 and #2 from Rano Raraku, according to raw data from Mann et al. (2008). The plot has been represented as in Fig. 4.12 for comparison. The height of the bars represents the dating error.

storms, and the corresponding changes in moisture balance, on Easter Island (Section 1.2). This was the first local and palynologically independent evidence of prehistoric climate change on Easter Island.

Pollen analysis of core #1 recorded the forest clearing at Raraku, but this was not considered to be caused by the above mentioned climatic drought, as palm decline began just after the hiatus. Before this sedimentary gap, the vegetation was totally dominated by palm forests, with sedges likely growing locally in the floating mat. Other trees, shrubs, and herbs were minor vegetation components. The situation was similar just after the hiatus but shortly after palms declined abruptly, whereas herbs such as grasses, sedges, *Polygonum*, and *Solanum* assumed predominance. This striking vegetation change coincided with an abrupt rise in sedimentary charcoal (Fig. 6.2) and

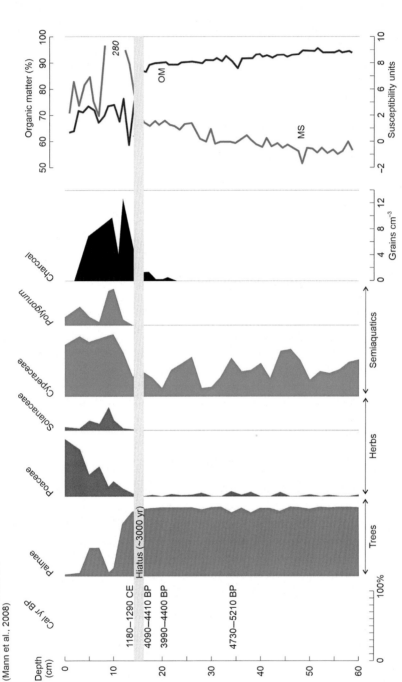

Fig. 6.2 Simplified percentage pollen diagram of core #1 from Rano Raraku (Mann et al., 2008). Semiaquatic plants were not included in the pollen sum used to calculate percentages. The depositional hiatus is indicated by a *gray band*. MS, magnetic susceptibility; OM, organic matter. *(Redrawn and modified from Mann, D., Edwards, J., Chase, J., Beck, W., Reanier, R., Mass, M., Finney, B., Loret, J., 2008. Drought, vegetation change, and human history on Rapa Nui (Isla de Pascua, Easter Island). Quat. Res. 69, 16–28.)*

also with the already mentioned sharp increase in magnetic susceptibility and the abrupt impoverishment in sedimentary organic matter (Fig. 5.4). Mann et al. (2008) also noted the chronological coincidence of forest clearing, as observed in the Raraku core, with their former study of primeval soils, whose erosion and further burial began by 1200 CE (Fig. 5.3). The authors highlighted that the observed sequence of events (rapid charcoal increases, widespread soil erosion, and increases of grasses and sedges) was typical and well documented in other Pacific islands after forest clearance by Polynesians. Therefore, they concluded that the agricultural colonization of Easter Island occurred in 1200 CE and was accompanied by rapid and widespread deforestation. Mann et al. (2008) pointed out that environmental transformation was very rapid and the existing stratigraphic records were probably too coarse to document the process, especially during the last 500 years. These authors recommended higher resolution studies to clarify the timing and patterns of the deforestation process, as well as the potential influence of eventual droughts that occurred after the depositional hiatus (1180–1290 CE onward).

The occurrence of an extensive depositional gap across Rano Raraku sediments was confirmed by a further systematic study based on a transect of eight new cores (RAR01 to RAR08) retrieved in 2006 with a UWITEC piston corer, combined with those from older studies (Fig. 3.6; Table 3.1) (Sáez et al., 2009). The sedimentary gap was first noticed in a composite section using the combination of four cores (Fig. 6.3) and then recognized across the whole transect, as a conspicuous and widespread erosional unconformity between sedimentary units 3 (peats deposited in a *Scirpus*-rich swamp) and 4 (more clastic sediments deposited after the recent lake refilling) (Fig. 3.11). The temporal extent of this sedimentary gap varied between approximately 5.8–4.2 cal kyr BP and 850–500 cal yr BP (1100–1450 CE). Sáez et al. (2009) related this hiatus with a mid-Holocene aridity crisis documented elsewhere in the circum-Pacific area, as a consequence of an insolation minimum leading to the weakening of the summer monsoon or, in agreement with Mann et al. (2008), a southward shift of storm tracks forced by El Niño-like dominant conditions. Sáez et al. (2009) also noted that the drought ended during the MCA, which was followed by a wetter phase, as suggested by the replenishment of the lake, co-inciding with the onset of the LIA. These authors used the argument of Nunn (2000) and Nunn and Britton (2001) that Easter island would have been colonized during the MCA, when warm climates and higher sea levels would have favored Polynesian navigation (Section 3.2). Therefore, Polynesian settlement and deforestation of Easter Island could have started before 850 cal yr BP (1100

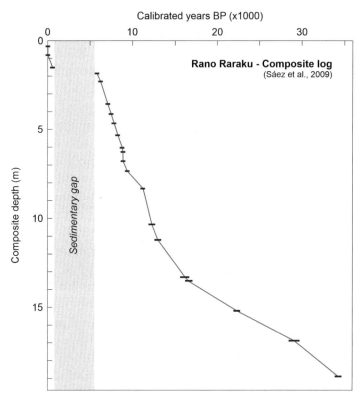

Fig. 6.3 Composite age-depth log using cores RAR01, RAR02, RAR03, and RAR07. *Red* bars are calibrated ages (2σ) after radiocarbon dating of *Scirpus* fragments and pollen-enriched extracts (Sáez et al., 2009). Six samples (not represented) were considered to be anomalous due to contamination from living roots (3) or by older material from lake margins (3). Calibration was performed with INTCAL98 (Reimer et al., 2004). *(Redrawn and modified from Sáez, A., Valero-Garcés, B., Giralt, S., Moreno, A., Bao, R., Pueyo, J.J., Hernández, A., Casas, D., 2009. Glacial to Holocene climate changes in the SE Pacific. The Raraku Lake sedimentary record (Easter Island, 27°S). Quat. Sci. Rev. 28, 2743–2759.)*

CE), during the mid-late Holocene drought, when there was no sedimentary record (Sáez et al., 2009). Although the authors did not provide evidence for these proposals, they emphasized that high-resolution pollen studies were in progress in the posterosive interval (850 cal yr BP onward) to test these ideas. The results of these palynological studies will be discussed in Section 6.2.2.

6.1.2 Rano Kao

In 2005, a new Rano Kao core (KAO05-3A) (Fig. 3.6; Table 3.1) was retrieved to record moisture variations and their potential influence on palm forest disappearance (Gossen, 2007). The core was at about 400 m from the lake shore, in a rather central position, and encompassed 2m of floating mat and 9m of bottom sediments, separated by about 10.5m of water. The floating mat recorded the last 605 ± 30 ^{14}C yr BP, calibrated to 1300–1400 CE onward, whereas the bottom sediments accounted for the time interval between $11,017\pm45$ and 1200 ± 35 ^{14}C yr BP, calibrated to about 11,100–10,900 and 770–900 cal yr BP (Fig. 6.4). All radiocarbon dates were obtained from *Scirpus* seeds and calibration was performed using the IntCal04 database (Reimer et al., 2004). Interestingly, the transition from the floating mat and the bottom sediments showed a chronological continuity, which contrasts with the former study of Flenley (1996) in KAO2, where the upper part of bottom sediments seemed to duplicate the floating mat record (Section 5.1). It is also noteworthy that the age-depth model of Gossen (2007) displayed a general chronological coherency

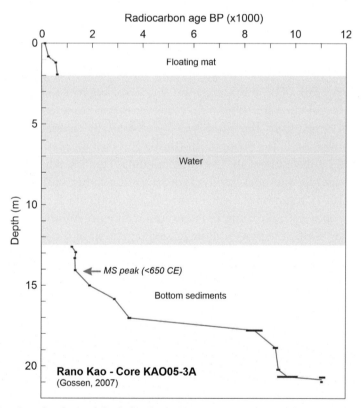

Fig. 6.4 Age-depth model of the Rano Kao core KAO05-3A after raw data from Gossen (2007). The length of the *red* bars represents the dating error and the *blue* line is the model used by Gossen (2007). The *green* arrow indicates the peak of magnetic susceptibility (MS) recorded before 650 CE.

(Fig. 6.4), which was absent in previous Rano Kao studies (Butler and Flenley, 2001; Butler et al., 2004), where age inversions were frequent. This suggested that totora seeds were likely the best material for radiocarbon dating of Rano Kao sediments, as Gossen (2007) emphatically expressed in the title of her paper: "The mystery lies in the *Scirpus*." A sharp increase of magnetic susceptibility was recorded by this author in the uppermost part of the bottom sediment, slightly below a radiocarbon date of 1335 ± 30 ^{14}C yr BP (650–720 CE). This event was followed by a significant increase in sedimentation rates (about two meters in only 40 years), which was interpreted as a sudden increase of soil erosion, possibly linked to deforestation. Further paleoclimatic and paleoecological studies were developed on core KAO05-3A using a variety of techniques including pollen analysis (Gossen, 2011). Unfortunately, to the knowledge of the author, this work remains unpublished.

Butler and Flenley (2010) resumed the analysis of their KAO2 core, with the aim of improving the chaotic age-depth model hitherto available (Section 5.1). That time, the authors avoided the use of plant fragments and pollen concentrates, as in former trials (Butler et al., 2004), and utilized a combination of the already available bulk samples and new ages obtained from *Scirpus* fruits. The chronology was somewhat improved but age inversions were still common (Fig. 6.5). However, when considering only the ages from *Scirpus* fruits, the model was chronologically coherent, with only one age inversion in the floating mat (Butler and Flenley, 2010). Two points not noted by the authors are worth mentioning. The first point is that the resulting age-depth model based on totora fruits is rather similar to that of Gossen (2007), based on the same material (Fig. 6.4), which supported the suitability of seed/fruit dating in Rano Kao sediments. Another interesting observation is that the previously suggested repetition of the mat sequence in the uppermost bottom sediments (Butler and Flenley, 2001; Section 5.1) was not observed in this case and there was an apparent chronological continuity between the base of the floating mat and the top of bottom lake sediments (Fig. 6.5).

Using this new chronology based on *Scirpus* fruits, Butler and Flenley (2010) subdivided the pollen diagram into 8 zones (Fig. 6.6), of which only that corresponding to the last millennia (RK2-4 to RK2-8) will be discussed here. According to Butler and Flenley (2010), the first good evidence of forest disturbance occurred at RK2-4, between 1900 and 1850 cal yr BP (50–100 CE), when the palm-tree forest was drastically reduced and replaced by grasses and ferns. This occurred after a conspicuous fire event, and Butler and Flenley (2010) speculate that this deforestation event might

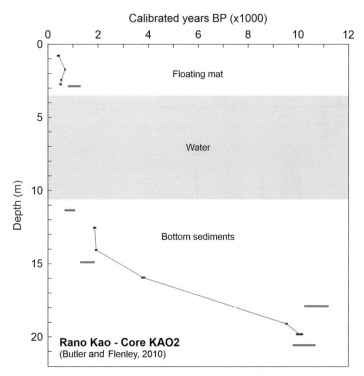

Fig. 6.5 Age-depth plot of the Rano Kao core KAO2, according to raw data from Butler and Flenley (2010). *Red* bars correspond to radiocarbon ages obtained on *Scirpus* fruits and *green* bars are ages from bulk sediment. The length of the bars is the 1σ calibration interval (the calibration method was not specified). The *blue* line is the age-depth model based on only *Scirpus* fruits.

have been caused by fires of volcanic, climatic, or anthropogenic origin. Forests partially recovered in RK2-5 (1850–1000 cal yr BP or 100–950 CE), probably due to an increase in precipitation, and experienced a further retreat at the end of the zone, also linked to a fire event. The presence of Urticaceae/Moraceae pollen was interpreted in terms of paper mulberry (*Broussonetia papyrifera*) cultivation. A further forest recovery was recorded in zone RK2-6 (1000–600 cal yr BP; 950–1350 CE), probably due to the migration of human populations out of the Kao crater. The major phase of final deforestation took place in RK2-7 (600–150 cal yr BP; 1350–1800 CE), and was clearly associated with an abrupt and extended increase of fire. Paper mulberry cultivation continued. The final extinction of palm trees occurred in RK2-8 (150 cal yr BP, or 1800 CE, to present). For Butler and Flenley (2010), the most interesting feature of the KAO2 diagram was the

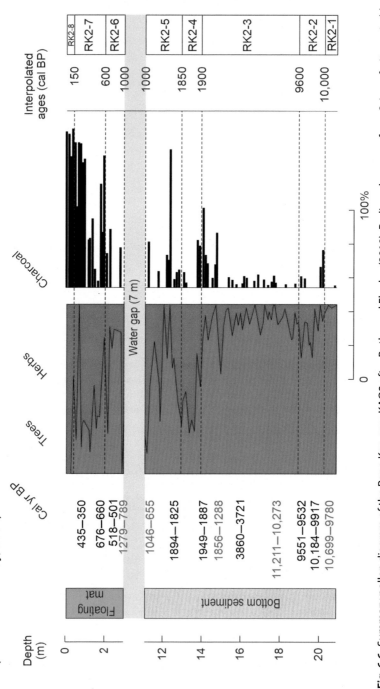

Fig. 6.6 Summary pollen diagram of the Rano Kao core KAO2 after Butler and Flenley (2010). Radiocarbon ages from *Scirpus* fruits are in *black* and those from bulk sediment are in *blue*.

possibility that human disturbance began around 100 CE and continued thereafter, perhaps varying in intensity but never ceasing. According to these authors, this interpretation would contradict most archaeological reconstructions and more research was needed for a sound assessment.

6.1.3 Soil investigations

Coinciding with the phase of lake coring intensification, Mieth and Bork (2010) extended their edaphic studies initiated on the Poike peninsula (Mieth et al., 2002; Mieth and Bork, 2003, 2005) (Section 5.2.2) to different parts of the island, mainly Rano Kao, Rano Raraku, and Maunga Orito (Fig. 3.12). The main aim of these studies was to test the hypotheses of humans, climate, or rats as the main deforestation agents. For this, the authors conducted physical, chemical, and biological analyses of soils and sediments from several exposures created by erosion and mining. Using the density and distribution of palm root casts on primeval soils, these authors estimated that approximately 16 million palm trees once grew on the island covering about 70% of its surface, with few palms growing above 250 m elevation in the Terevaka highs. The same authors later extended their estimates to 20 million palms covering 80% of the island (Mieth and Bork, 2015, 2017; Section 2.4.4). Similar to Mann et al. (2003), Mieth and Bork (2010) found evidence of slash and burn in the form of charcoal layers, often containing charred palm nutshells, situated above the level of the uppermost palm root casts (Fig. 6.7). Radiocarbon dating of these charcoal layers at different points of the island yielded dates of 1200 CE and 1500 CE for the beginning and the end, respectively, of intensive slash and burn practices, which is also in agreement with previous studies by Mann et al. (2003) using similar techniques. Remains of human activity and stone constructions were directly on top of burned surfaces, supporting extensive anthropogenic forest clearing.

Mieth and Bork (2010) found evidence of palm woodland regeneration in the inner slopes of the Raraku crater. There, a primeval soil with palm root casts dating back to 1283–1394 CE was covered by a colluvial layer deposited after the first clearance, as usual. But this colluvial layer contained traces of a second generation of younger palms in the form of root casts, which were interpreted as evidence of woodland regeneration after local forest clearing. A second deforestation event occurred by 1402–1436 CE definitively removed the palm woodland from the site. The authors also found that less than 10% of the charred palm nuts showed teeth marks from rats and concluded that these animals could have played a minor role in inhibiting forest regeneration but they could not have been the main deforestation agent, as previously suggested

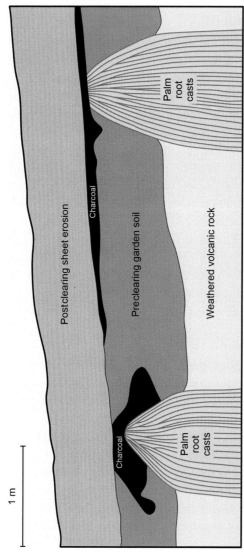

Fig. 6.7 Soil profile from Poike showing the upward succession of a primeval soil with weathered volcanic rocks and garden soils with palm root casts. Forest clearing left some charcoal accumulations on the top of the primeval soil that was truncated by erosion and covered by a sheet of fine-layered sediments. *(Redrawn and modified from Mieth, A., Bork, H.R., 2010. Humans, climate or introduced rats—which is to blame for the woodland destruction on prehistoric Rapa Nui (Easter Island)? J. Archaeol. Sci. 37, 417–426.)*

by Hunt (2006, 2007) (Section 2.4.4). Mieth and Bork (2010) concluded that the island-wide forest clearing was exclusively human-made. To those dates, paleoecological evidence pointed toward a single and relatively rapid island-wide event of forest clearing but the study of Mieth and Bork (2010) showed that Easter Island's palm woodland was able to regenerate after burning, which was consistent with the palynological record of core KAO2 from Rano Kao, where two forest recovery events were recorded, one between 100 and 950 CE and another between 950 and 1350, prior to the final extinction of palms, which did not occur until about 1800 CE (Butler and Flenley, 2010).

The frequent occurrence of palm root casts and agricultural soils together (Fig. 6.7) supported the results of previous studies, developed at Te Niu (Fig. 3.12), which suggested the occurrence of disturbed forests and a mixed crop dryland production system. Indeed, Wozniak et al. (2007) and Horrocks and Wozniak (2008), using pollen, phytolith, and starch analysis from archaeological sites, recorded the joint occurrence of altered forests and crops of bottle gourd (*Lagenaria siceraria*), yam (*Dioscorea alata*), sweet potato (*Ipomoea batatas*), and taro (*Colocasia esculenta*) before deforestation, which occurred at about 1300 CE. These studies were further extended to the whole island (Horrocks et al., 2016) and will be discussed in Section 6.2.1.

6.2 Publication resurgence

At the beginning of the present decade (2010–2020), the legacy of more than 30 years of paleoecological research on Easter Island was enormous, in comparison with former times. However, what happened during the last millennia, which is crucial for a proper understanding of the island's prehistory in terms of ecological and environmental change, continued to be hidden by a sedimentary gap (Rano Raraku), the human disturbance of the most recent sediments (Rano Aroi) and the lack of a reliable high-resolution age-depth model of the most recent sedimentary record (Rano Kao). But the renewed coring efforts of the 2005– 2009 period (Fig. 3.7) began to bear fruit some years later in the form of papers reconstructing the main paleoecological trends using mainly microfossil analyses, notably palynology, but also a variety of physicochemical proxies. Here we will subdivide these advances, into two conceptual groups. The first category encompasses studies based on biological analyses (notably pollen, phytoliths, and starch) on cores from the three ranos and an island-wide soil survey, focused on the reconstruction of prehistoric cropping activity. The second category gathers multiproxy (physical, chemical, and biological) studies on continuous, mostly gap-free, cores from the same waterbodies, aimed at reconstructing the spatiotemporal deforestation patterns and their potential climatic and/or anthropogenic drivers.

6.2.1 Polynesian crops

The first records of Polynesian cultigens were based on pollen, phytoliths, and starch analysis in agricultural and garden soils from the archaeological site of Te Niu (Section 6.1.3). The only pollen type linked to a cultivated plant found on lake sediments was from Rano Kao and corresponded to the Urticaceae/ Moraceae type, possibly *Broussonetia papyrifera* (paper mulberry), identified by Butler and Flenley (2010) in Core KAO2 (Section 6.1.2). To improve this situation, Horrocks et al. (2012a,b) used, for the first time, pollen, phytolith, and starch analysis on Easter Island lake sediments and essayed a new method, based on Fourier Transform Infrared Spectroscopy (FTIR), to identify degraded sedimentary starch particles (Higgins et al., 1961; Goodfellow and Wilson, 1990). These analyses were performed in a transect of four cores (2–4) retrieved along the SW margin of Rano Kao, just below the ceremonial village of Orongo (Fig. 3.6), the center of the Birdman cult (Section 2.4.3). The authors noted that, in addition to Orongo, a heavy concentration of archaeological sites was recorded around Rano Kao (Ferdon, 1961; McCoy, 1976; Vargas et al., 2006) (Fig. 1.16). Inside the crater, terracing for gardens and dwellings was common. Although using totora fruit/seeds, radiocarbon dating of the four cores retrieved showed, once more, frequent age inversions, which prevented the attainment of a reliable chronology (Fig. 6.8). According to Horrocks et al. (2012b), this chronological disorder was due to sediment mixing as a result of human activity. In spite of this, microfossil analysis was successful and, as previously reported for some archaeological sites (Horrocks and Wozniak, 2008), the scenario was consistent with a landscape of large-scale deforestation (as recorded by phytoliths) and a mixed-crop production system including common Polynesian cultigens such as paper mulberry, taro, yam, sweet potato, bottle gourd, and banana (*Musa* sp.), which were identified mainly from starch remains. This study demonstrated that the SW margin of Lake Kao was an actively cultivated area during and after its deforestation, likely favored by the continuous availability of water and the protection from the dominant winds (Horrocks et al., 2012b).

A further core (1) in the middle of the former transect but in a less marginal position (Fig. 3.6) was obtained by the same research team and analyzed for pollen, phytoliths, and ant (Formicidae) remains (Horrocks et al., 2013). The base of the core provided the oldest ages known to date for Rano Kao (34,260 ± 440 and 20,340 ± 160 ^{14}C yr BP) followed by a

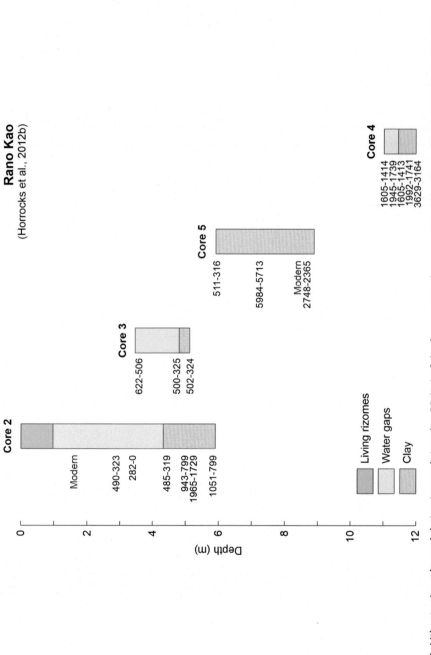

Fig. 6.8 Lithostratigraphy and dating intervals in cal yr BP (2σ) of the four marginal cores retrieved by Horrocks et al. (2012b) in the SW margin of Rano Kao, below the ceremonial village of Orongo. Calibration followed McCormac et al. (2004) and Reimer et al. (2004). Water gaps included some detritic levels with enough organic matter for radiocarbon dating.

thin Late Glacial layer (13,210–12,110 cal yr BP) and a sedimentary gap between this date and approximately 3500 cal yr BP. This hiatus coincided with an extended phase of extremely low sedimentation rates as recorded in core KAO05-3A (Gossen, 2007), possibly due to dryness and low lake levels (Fig. 6.4). This sedimentation disruption was used by Horrocks et al. (2013) to suggest a mechanism of formation of the floating mat that has been explained in Section 3.5 (Fig. 3.10). After 3500 cal yr BP, the record was continuous and chronologically coherent until 1000 cal yr BP. But this was only true for the bottom sediments below the water gap, as in the floating mat, a reliable chronology was still elusive (Fig. 6.9).

A surprising finding was the very high concentration of charcoal fragments and the presence of human-introduced *Musa* (banana) phytoliths at the base of the late Holocene diagram, between about 3680 and 2860 cal

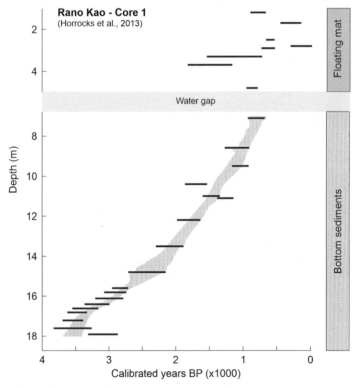

Fig. 6.9 Calibrated (2σ) radiocarbon dates *(red bars)* on totora seeds/fruits and Bacon age-depth modeling *(gray area)* for the Holocene interval Rano Kao core 1 (Horrocks et al., 2013). Calibration after Reimer et al. (2004) and age-depth modeling after Blaauw and Christen (2011).

Rano Kao - Core 1
(Horrocks et al., 2013)

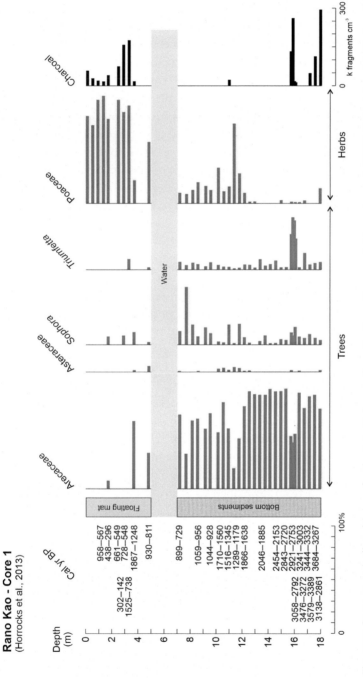

Fig. 6.10 Percentage pollen diagram for the Late Holocene from the Rano Kao core 1 (Horrocks et al., 2013). Only dates obtained on totora seeds/fruits are indicated (see Fig. 6.9 for more dating details). *(Redrawn and modified from Horrocks, M., Marra, M., Baisden, W.T., Flenley, J., Feek, D., González-Nualart, L., Haoa-Cardinali, S., Edmunds Gorman, T., 2013. Pollen, phytoliths, arthropods and high-resolution 14C sampling from Rano Kau, Easter Island: evidence for late quaternary environments, ant (formicidae) distributions and human activity. J. Paleolimnol. 50, 417–432.)*

yr BP (1730–910 BCE) (Fig. 6.10). According to Horrocks et al. (2013), this could be indicative of an erosion event as a result of forest clearance by fire for cultivation purposes. However, this would have occurred roughly a couple of millennia prior to the generally accepted dates of settlement of Eastern Polynesia (Section 2.3.2). After a short charcoal-free interval, a second charcoal peak was recorded at 3060–2750 cal yr BP (1110–800 BCE), also coinciding with the presence of banana phytoliths, which reinforced former observations. In this case, a significant increase of *Triumfetta* at the expense of palms (Arecaceae) was interpreted as a successional forest response, or as an evidence of cultivation of *Triumfetta* for rope manufacturing (Horrocks et al., 2013). The authors discussed the possibility of these dates to be older than expected due to the hard water effect (Olsson, 1986) or to the mixing of ^{14}C-dead CO_2 with aquatic or atmospheric CO_2, due to local volcanic activity (González-Ferrán et al., 2004). A third even of possible Polynesian arrival was recorded at 1289–1179 cal yr BP (661–771 CE), which was more consistent with the dates generally accepted for Easter Island settlement (Section 2.3.2). This event was characterized by a sharp decline of Arecaceae, a significant peak of Poaceae, and the presence of some charcoal (Fig. 6.10). The proliferation, at the same level, of midges (Chironomidae), shore flies (Ephydridae), and mites (Oribatidae), and the presence of introduced ants (Formicidae) was considered by the authors as an additional evidence of human disturbance. Despite inconsistent dating, the floating mat showed clear evidence of major landscape disturbance by people, from 4 m to the top of the core (Horrocks et al., 2013). Notable are the disappearance of palms, Asteraceae, and *Triumfetta* along with the large amounts of charcoal, which indicated large-scale clearance of the remaining forests by fire. The resulting slope erosion could have introduced older carbon to the sediments, thus contributing to age inversions.

The same research team developed extensive studies within the Rano Raraku crater aimed at unraveling whether the site was used as a moai quarry alone or was also cultivated during Easter Island's prehistory (Horrocks et al., 2012a). For this, the authors analyzed macrofossils of dryland plants and conducted combined microfossil analysis (pollen, phytoliths, and starch) in a lake core and also in the surrounding dryland soils. The finding of a consistent alignment of mineralized (iron oxide) root casts attributed to *Scirpus californicus* some 10 m above the present lake level was remarkable, as it suggested that the floating mat, and therefore the water level, was 10 m above the present sometime in the past. This coincided with earlier observations by Flenley et al. (1991) and represented the maximum flooding possible for the Raraku catchment, which probably overflowed by its SW side (Fig. 3.6). Horrocks et al. (2012a) suggested that this could have occurred prior to 28,000 cal yr BP when, according to Sáez et al. (2009), Lake Raraku attained water levels much higher than today. An unnamed sediment core, tentatively designed here as "Lake," obtained on the eastern side of the lake (Fig. 3.6; Table 3.1) contained several age inversions and a sedimentary gap of about 8000 years (10,179–9787 to 1862–1635 cal yr BP) between 0.9 and 1.0 m (Fig. 6.11). Horrocks et al. (2012a) argued that this could be due to the location of the coring site near the shore (Fig. 3.6), where the danger of aerial exposure and the resulting erosion was higher than in other, more central localities.

In this case, the interval of interest is the time after the hiatus, which corresponds to the uppermost meter of sediments formed mainly by clay. The abrupt replacement of palm forests by grass meadows occurred at 625–513 cal yr BP (1325–1437 CE) (Fig. 6.12), coinciding with a significant fire exacerbation and the first appearance of starch of sweet potato

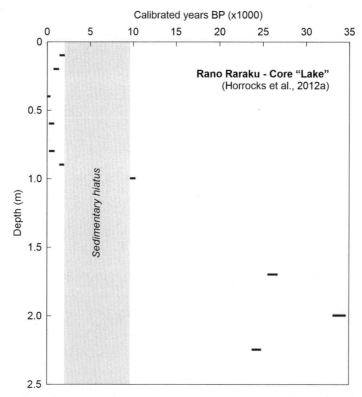

Fig. 6.11 Age-depth plot for the Rano Raraku lake core of Horrocks et al. (2012a). Dates were obtained on *Scirpus* fruits/seeds and calibrated according to McCormac et al. (2004) and Reimer et al. (2009).

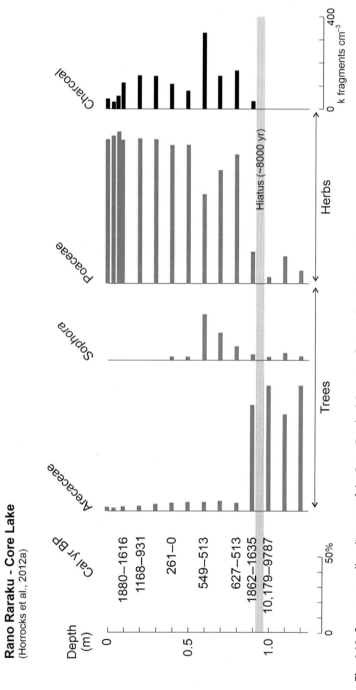

Fig. 6.12 Summary pollen diagram of the Rano Raraku lake core of Horrocks et al. (2012a).

(*Ipomoea batatas*) and taro (*Colocasia esculenta*). The subsequent increase of *Sophora* could be due to its ability to fill gaps during forest clearance, and its sudden decline after 549–513 cal yr BP (1401–1437) was considered to be the results of the almost complete deforestation of the catchment (Horrocks et al., 2012a). A dryland soil profile, whose basal date was 1174–1050 cal yr BP (776–900 CE), was totally dominated by grasses and was interpreted as a postclearing unit, which was in conflict with the date of deforestation measured in the lake core. The authors suggested that the soil profile could be the result of episodic deposition resulting from upslope erosion and, hence, it could contain strata with carbon older than the onset of anthropogenic deforestation. Palm phytoliths were also abundant in the central part of the soil profile when, according to pollen studies, palms were no longer growing on the site. The explanation provided by Cummings (1998) in her pioneering phytolith studies could be invoked here. Indeed, palm phytoliths may be abundant even after deforestation as they are used to concentrate on soils after the decay of the organic parts of dead palms (Section 5.4). Horrocks et al. (2012a) concluded that the Rano Raraku catchment and its surroundings were used not only as a moai quarry but also as an extensive multicropping site, and suggested that the crater was intensively gardened and terraced, possibly during the peak of the moai-quarrying period.

Horrocks et al. (2015) also cored Rano Aroi (core RA) and a nearby small depression called Rano Aroi Iti (core RAI) using a Russian corer (Fig. 3.6; Table 3.1). They also studied some surrounding soil profiles from terraces and dry deposition sites, as well as a cave. The main aim of that study was to evaluate the timing and nature of deforestation and agricultural activity in the highest elevations of the island, in comparison with lowland sites. The study was focused on the last millennium and included pollen, phytoliths, starch, diatoms, and arthropods. Radiocarbon dating was performed on *Scirpus* seeds/fruits and calibrated using the SHCal13 database, especially developed for the Southern Hemisphere (Hogg et al., 2013). Age-depth modeling used the Bacon Bayesian approach (Blaauw and Christen, 2011). The age-depth model for core RA resulted to be poorly constrained from 12,000 to 1000 cal yr BP. A set of dates were rejected for being older or younger than expected, probably due to vertical movements of macrofossils, contamination, overturning of floating mats, or volcanic CO_2 influence (Horrocks et al., 2015). The model was well constrained for the upper 1.2 m, containing the last 1500 cal yr BP, and was therefore appropriate to analyze evidence associated with human activity (Fig. 6.13). The Rano Aroi Iti core (RAI) was 2 m deep and contained a record of the last 1500 years.

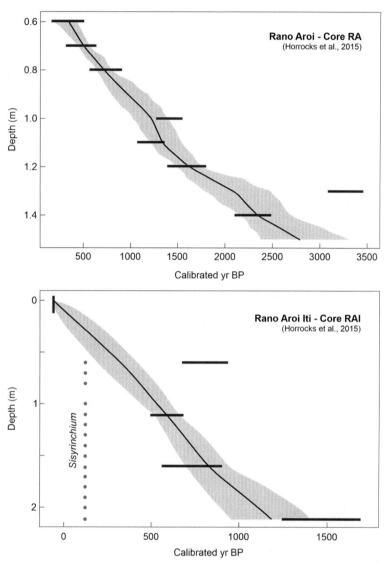

Fig. 6.13 Bayesian age-depth models for the upper 1.5 m of the Rano Aroi core RA (above) and the whole core RAI (below) (Horrocks et al., 2015) using the Bacon method (Blaauw and Christen, 2011). *Red* bars represented the calibrated dates (2σ), the *black* solid line is the model's best fit, and the *gray* area is the 95% confidence interval. *Blue* dots in the core RAI are samples with the presence of *Sisyrinchium*.

According to Horrocks et al. (2015), the sudden increase of charcoal and the total disappearance of pollen of Arcecaeae and all woody taxa in core RA (Fig. 6.14) clearly reflected large-scale anthropogenic deforestation by burning. This event commenced at 0.8 m depth that, according to the Bacon age-depth model (Fig. 6.13), corresponded to 709 (645–797) cal yr BP or 1241 (1153–1305) CE. Total forest clearance at Rano Aroi occurred shortly after, at 0.6 cm, corresponding to an age of 339 (177–428) cal yr BP or 1611 (1522–1773) CE. The top 0.6 m of the core corresponded to the postcontact period and reflected modern agroforestry (*Pinus*, Myrtaceae), as well as the formation of the modern swamp vegetation, which is dominated by *Scirpus* and *Polygonum*. An interesting point was the presence, just before total deforestation, of pollen from *Sisyrinchium*, probably *S. michrantum*, and invasive American weed widely naturalized elsewhere. In core RA, this pollen type was present in a single sample but in core RAI, *Sisyrinchium* occurred in a consistent manner in almost all the samples examined (Fig. 6.13), since at least 1542–1316 cal yr BP (408–634 CE). The authors noted the chronological incompatibility of this observation with the arrival of the first Europeans (1722 CE) and mentioned a previous finding by Cañellas-Boltà et al. (2012) in Rano Raraku that could suggest an early Amerindian contact long before the Polynesian arrival (Section 6.2.2). Other possibilities suggested by Horrocks et al. (2015) were that the Rano Aroi Iti would have been disturbed to its full depth, thus giving ages older than expected, or that the whole sequence was formed after the European contact.

Evidence for horticultural activity was limited in the swamp cores, as only a few starch grains of cf. *Colocasia esculenta* were found in a few samples of core RA. Some terraces and the cave provide additional evidence in the form of phytoliths of *Musa* and cf. *Broussonetia* (Horrocks et al., 2015). These authors noted that the evidence of horticulture was similar to that identified at lowland sites (Horrocks et al., 2012a,b, 2013) except for the lack of remains of *Ipomoea batatas* in Rano Aroi, where it seems not to have been an important local cultigen. According to some radiocarbon dates from the cave and a terrace, the authors suggested that agricultural activity in Rano Aroi took place either between 1670 and 1740 CE or between 1795 and the introduction of sheep farming (Section 1.4.3). In either case, Rano Aroi seems to have been occupied by humans in a more or less continuous fashion later than lowland sites (Rao Kao and Rano Raraku).

The study of prehistoric agricultural activities using microfossils (pollen, starch, and phytoliths) was completed with an extensive survey of dryland soils across the island (Fig. 3.12), carried out by the same team (Horrocks et al., 2016), aimed at determining the success and extent of Easter Island's agriculture in the

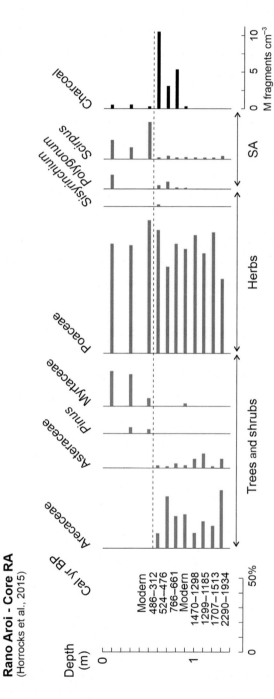

Fig. 6.14 Summary pollen diagram of the upper 1.5 m of core RA from Rano Aroi (Horrocks et al., 2015). The dotted line represents the post-contact zone. SA, semiaquatics.

past. Of the 11 sites analyzed (15 soil profiles and 40 samples in total), only one soil profile was dated (TV1 from the Terevaka), using undifferentiated fruits/seeds and the SHCal13 calibration curve (Hogg et al., 2013). The two dates obtained were "modern" at 5–10 cm and 1944–1735 cal yr BP (6–215 CE) at 60–65 cm. The authors considered that the last date could be older than expected by remobilization of plant material as a result of human activities. They also inferred that other profiles broadly followed the same stratigraphic patterns and considered that the soils sampled corresponded to times after Polynesian settlement. These analyses confirmed that the Polynesian cultigens previously identified in the three freshwater catchments (Rano Aroi, Rano Kao, and Rano Raraku) and the archaeological site of Te Niu were cultivated across most of the island. Horrocks et al. (2016) identified *Brousonetia papyrifera* (paper mulberry), a species used by the ancient Rapanui to make bark clothing, using pollen and phytoliths. This taxon was found in 45% of the sites including the higher elevations of the island. *Musa* (banana) phytoliths were found only in one site. The fruit of these plants was for human consumption and the leaves were utilized for house covering and tying purposes in house and garden construction. The identification of *Ipomoea batatas* (sweet potato) starch was problematic because of its similarity to other cultivated species and the indigenous *Ipomoea pes-caprae*. Horrocks et al. (2016) considered that the starch found in 36% of their dryland soils corresponded to sweet potato. Pollen and starch grains of *Colocasia esculenta* (taro) were found at 55% of the sites, including high-elevation localities.

6.2.2 Continuous deforestation records

Parallel to Horrocks' team publications, our own paleoecological team published the results of pollen analyses on cores retrieved during the 2006 and 2008 campaigns (Table 3.1), where a set of continuous and quasicontinuous, and chronologically coherent—that is, free from large sedimentary gaps and age inversions—records of the last millennia were obtained in Rano Aroi, Rano Raraku, and Rano Kao (Rull et al., 2013, 2016, 2018). In addition, some of the studies were multiproxy surveys and included nonbiological proxies that provided independent paleoclimatic evidence, thus avoiding circularity. Other longer records were published encompassing the last 70,000 years (Cañellas-Boltà et al., 2012, 2014, 2016; Margalef et al., 2013, 2014), but only those containing the last few millennia will be discussed here. These records corresponded to cores RAR08 for Raraku, ARO 08-02 for Aroi and KAO08-03 for Kao (Fig. 3.6; Table 3.1). The continuity and chronological coherency of the age-depth models for these three cores can be seen in Fig. 6.15. Only in Rano Kao, a date of 5060 ± 20 ^{14}C yr BP at 247 cm had to be rejected.

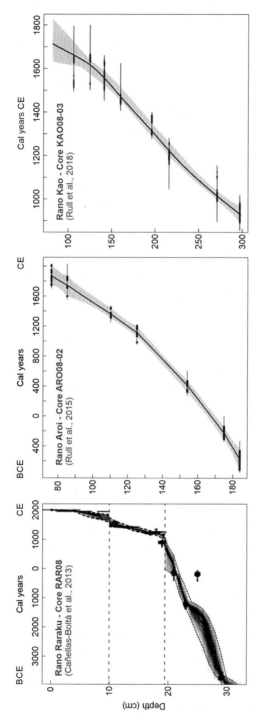

Fig. 6.15 Age-depth models of the continuous and chronologically coherent sedimentary records of the last millennia recovered in Rano Raraku, Rano Aroi, and Rano Kao. All radiocarbon dates were obtained from pollen extracts (Vandergoes and Prior, 2003) and calibrated using SHCal13 (Hogg et al., 2013). The Rano Kao model was obtained using the Bayesian Bacon method (Blaauw and Christen, 2011) and the others were constructed using local interpolation (Rano Aroi) and the smooth spline method (Rano Kao) (Blaauw, 2010). Dotted lines in the Raraku model around 10 and 20 cm indicate minor sedimentary gaps (see text for explanation). *(Redrawn and modified from Rull, V., 2020. The deforestation of easter Island. Biol. Rev. 95, 121–141.)*

The Raraku core (RAR08) included the last five millennia (Fig. 6.15), which were mostly missing in former works (Flenley et al., 1991; Dumont et al., 1998; Mann et al., 2008; Sáez et al., 2009) due to extensive, millennial-scale depositional gaps. In this core, only two minor sedimentary gaps were detected by the Bacon model between about 500 and 1165 CE (665 year duration), and between 1570 and 1720 CE (150 year duration), which were interpreted as the consequence of depositional cessation and/or increased erosion during dry phases (droughts) that likely resulted in the total drying out of the lake (Cañellas-Boltà et al., 2013). The first drought ended during the MCA and the second took place during the LIA. The pollen record showed that, contrarily to former interpretations, deforestation did not occur in an abrupt manner. Rather, palm forest clearing was gradual with three main decline pulses at 450 BCE, 1200 CE, and 1475 CE (Fig. 6.16). The first deforestation pulse (450 BCE) was accompanied by an increase of grasses and the first appearance of *Verbena litoralis*, coinciding also with an increase of charcoal. This constituted the first palynological record of *V. litoralis* in Easter Island cores (Fig. 6.17). The interesting aspect of this finding is that *V. litoralis*, a species native to tropical America and commonly associated with anthropogenically disturbed sites (agricultural or ruderal), had previously been considered to be introduced in Easter Island after European contact (Zizka, 1991). Cañellas-Boltà et al. (2013) discussed whether the earlier arrival of *Verbena* and its successful establishment on Easter Island could have been the result of a chance long-distance dispersal or of human arrival much earlier than formerly expected. According to the authors, the coincidence of the settlement of *Verbena* with the first palm decline and the first charcoal increase supported the second option, which would involve the human discovery of Easter Island some 1500 years before the generally accepted date. This would be consistent with the hypotheses about an assumed Amerindian arrival before Polynesian settlement (Section 2.3). It is also interesting to note that the first deforestation pulse occurred when the catchment was occupied by a mire instead of a lake, as indicated by the peaty sediments and the dominance of benthic diatoms, along with the absence of planktic species (Fig. 6.16). This suggested that climates were drier than at present but not arid.

The second deforestation pulse, which occurred shortly before 1200 CE, represented a major ecological disturbance characterized by a significant forest decline and an outstanding expansion of grass meadows, as well as of *Verbena* populations (Fig. 6.16). The abrupt increase of charcoal suggested that these events were linked to a remarkable fire exacerbation, likely of anthropogenic origin. These changes coincided with a significant shift in the diatom record, from benthic to planktic assemblages, which suggested a change from mire to

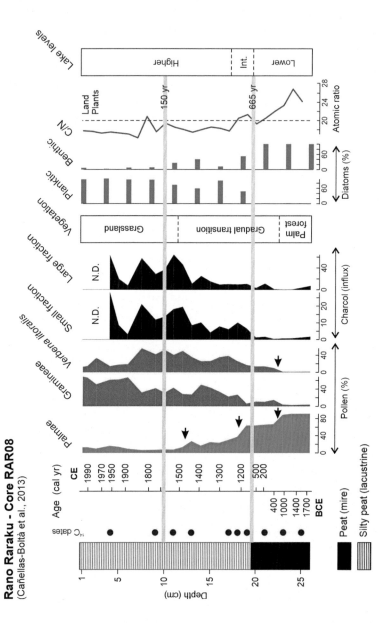

Fig. 6.16 Summary diagram of the last c.3700 years as represented in core RAR08 from Rano Raraku (Cañellas-Boltà et al., 2013). Sedimentary gaps are represented as gray bands with indication of the years lacking in each case. Black arrows indicate remarkable events discussed in the text. N.D., no data. *(Redrawn and modified from Cañellas-Boltà, N., Rull, V., Sáez, A., Margalef, O., Bao, R., Pla-Rabes, S., Blaauw, M., Valero-Garcés, B., Giralt, S., 2013. Vegetation changes and human settlement of Easter Island during the last millennia: a multiproxy study of the Lake Raraku sediments. Quat. Sci. Rev. 72, 36–48.)*

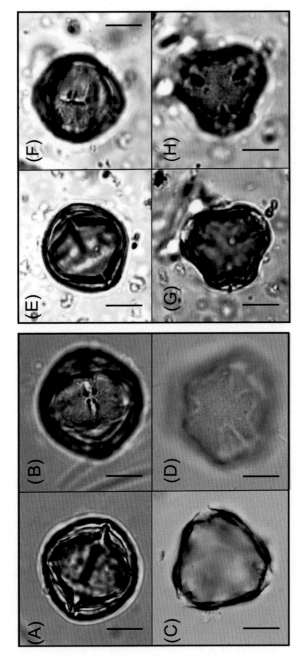

Fig. 6.17 *Verbena litoralis* pollen. (A)–(D), pollen from the species presently growing within the Raraku crater, near the coring site, collected in the field by N. Cañellas-Boltà. (E)–(H), pollen from sediments of core RAR08 (Caññellas-Boltà et al., 2013). Vertical bars represent 10 μm. *(Modified from Cañellas-Boltà, N., Rull, V., Sáez, A., Margalef, O., Bao, R., Pla-Rabes, S., Blaauw, M., Valero-Garcés, B., Giralt, S., 2013. Vegetation changes and human settlement of Easter Island during the last millennia: a multiproxy study of the Lake Raraku sediments. Quat. Sci. Rev. 72, 36–48.)*

lacustrine environments, likely due to the return of wetter climates after the MCA drought. The acceleration of fire incidence and forest clearing in such wetter climates reinforced that humans would have been the main agent responsible for landscape disturbance. The third forest clearing event (1475 CE) represented the total deforestation of the Raraku catchment and the irreversible establishment of grass meadows, also associated with a significant charcoal increase. Of the three recorded palm declines, only the second could possibly be linked to a climate change, as it occurred immediately after the first drought inferred from the depositional gap recorded between 500 and 1165 CE. Cañellas-Boltà et al. (2013) suggested that climate and human activities could have acted synergistically in the demise of forests. According to these authors, drought would have favored vegetation ignitability, thus facilitating forest burning. After the LIA drought, planktic diatom assemblages dominated and benthic species were absent, thus suggesting maximum lake levels and wetter climates.

The Rano Aroi core ARO08-02 provided a continuous record of the last 2600 years, with the exception of the last 2 centuries (Rull et al., 2015). Four main vegetation phases could be distinguished (Fig. 6.18). During Phase I (750 BCE–1250 CE), the site was covered by a palm woodland with Asterceae shrubs in the understory, within a regional landscape dominated by grass meadows. Semiaquatic plants typical of present-day swamp vegetation (Cyperaceae, *Polygonum*) and ferns were absent, suggesting that the swamp was probably dry or the water level was lower than today. In spite of the general uniformity of this phase, a recurrent pattern emerged in the form of successive palm, Asteraceae and grass peaks likely representing processes of landscape opening (LO) from palm-dominated to grass-dominated landscapes. The first of these events (LO1) occurred between 300 BCE and 50 CE, and the second (LO2) between 600 and 1100 CE (Fig. 6.18). Between 1250 and 1520 CE (Phase II), a progressive densification of palm woodland to more closed forests took place, along with a change in the shrubby understory, where *Sopohora* coexisted with Asteraceae. At the same time, the local semiaquatic vegetation appeared, which indicated an increase in the water table, probably due to wetter climates, which would have favored forest growth. During Phase III (1520–1620 CE), palm forests and the associated shrublands were abruptly removed and replaced by grass meadows, coinciding with a sudden charcoal increase, which suggested anthropogenic burning. The palynological expression of this deforestation event (LO3) was similar to former LO1 and LO2 events but was significantly faster and direct, from palms to grasses, without an intermediate shrub stage. Present-day vegetation, consisting of a continuous cover of grass meadows with no signs of the former forests and shrublands, was established after 1620

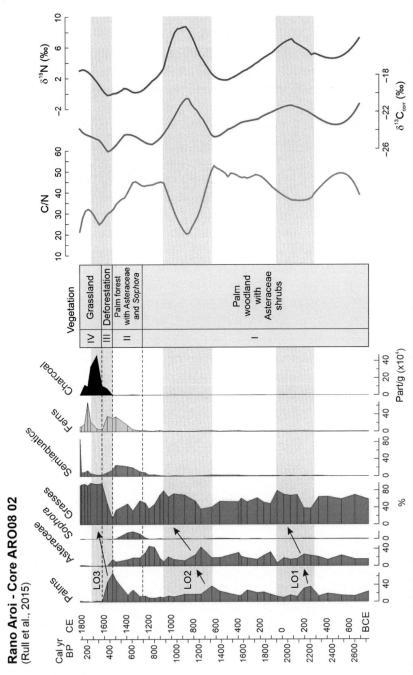

Fig. 6.18 Summary diagram of core ARO08-03 (Rano Aroi) showing the main palynomorphs and the main geochemical parameters used as paleoclimatic proxies (Rull et al., 2015). *Gray bands* indicate drier climates linked to processes of landscape opening (LO1, LO2, LO3).

CE (Phase IV). It is noteworthy that the charcoal peak occurred during this phase, when they were no longer woody elements growing in the site. The coincidence of the charcoal peak with the decline of semiaquatic plants and ferns would be compatible with burning of local swamp vegetation.

Some geochemical proxies underwent relevant shifts during LO1 and LO2 events, in the form of significant C/N declines (due to N increases) linked to conspicuous increases in $\delta^{15}N$ and $\delta^{13}C$ (Fig. 6.18). The increase in C/N, coupled with a decrease in $\delta^{15}N$, was compatible with lower water levels and eventual drying (Houlton and Bai, 2009; McLauchlan et al., 2013). The increase in $\delta^{13}C$ could have been due, in part, to the general increase of C_4 grasses to the detriment of woody C_3 plants and algae (O'Leary, 1988). Therefore, the geochemical signal was consistent with the occurrence of dry climates during the LO1 and LO2 events. The LO3 event was more difficult to interpret in climatic terms, as humans were likely involved in forest clearing. However, geochemical proxies showed a trend similar to the earlier part of LO1 and LO2 events, suggesting the onset of a shift toward drier climates (Rull et al., 2015). Interestingly, the event LO2 (600–1100 CE) was roughly coeval to the MCA Raraku drought (500–1165 CE) and the event LO3 (1520–1700 CE) occurred at similar times to the LIA Raraku drought (1570–1720 CE). This coincidence suggested the occurrence of generalized dry climates across Easter Island, although they were more intense in the lowlands (Raraku), where they caused the interruption of the sedimentation. It is possible that the increasing precipitation gradient with elevation (Section 1.2) could have attenuated the drought intensity in Aroi, where only vegetation changes, but no drying out, were recorded.

The Rano Kao core KAO08-03 recorded the time interval between about 950 and 1710 CE (Seco et al., 2019). The analysis of this core is still in progress to increase the resolution but the main vegetation trends over time have already been resolved. A novelty of this analysis with respect to former studies on Easter Island sediments was the identification of fungi spores linked to human activities, with emphasis on *Sporormiella*, a genus of coprophilous fungi growing in the dung of domestic vertebrates, and therefore an indirect indicator of human presence (Van Geel et al., 2003) (Section 2.3.1). According to Seco et al. (2019), the Kao catchment was covered by an open-palm woodland, rather than a closed-palm forest, around 1000 CE, which suggested that the Kao forests were already being disturbed. From that point, the deforestation was gradual but spiked by three acceleration pulses at about 1070 CE, 1410 CE, and 1600 CE (Fig. 6.19). Each of these pulses was followed by a regeneration trend except the third, after

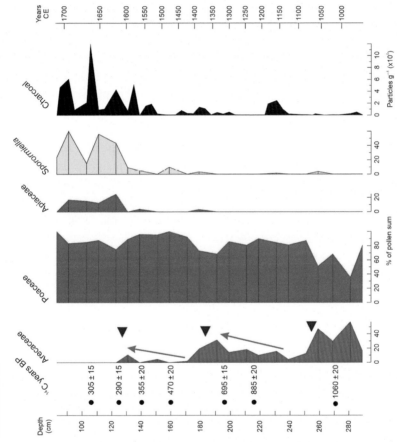

Fig. 6.19 Summary pollen diagram of the Rano Kao core KAO08-03 after Seco et al. (2019). Deforestation pulses are marked by *red* triangles and forest regeneration trends are indicated by *blue* arrows.

which the forests did not recover anymore. The first pulse of forest decline (1070 CE) occurred during the MCA drought recorded at Rano Raraku and Rano Aroi, which suggested a potential influence of climate. A brief fire event between about 1150 and 1200 CE did not affect general forest regeneration, possibly favored by wetter climates. The second forest decline (1410 CE) coincided with another charcoal peak and the first continuous appearances of *Sporormiella* spores, suggesting the onset of a continuous anthropogenic disturbance. Palms almost disappeared but recovered slightly until the third declining pulse (1600 CE), when charcoal and *Sporormiella* increased dramatically, roughly coinciding with the first phases of the LIA drought, which could have contributed to fire exacerbation. A general conclusion was that humans were present in the Rano Kao catchment since the beginning of the KAO08-03 record but in dispersed or occasional populations. Human presence became more or less permanent, but still scarce, by 1410 CE. The catchment was fully deforested and significantly disturbed by larger human populations at approximately 1600 CE, coinciding with the first activities in the ceremonial village of Orongo, the center of the Birdman cult (Robinson and Stevenson, 2017). At the same time, the pollen of Apiaceae, likely *Apium*, increased abruptly and remained constant until at least 1700 CE, which suggested cultivation of this plant, as already proposed by Flenley et al. (1991) in their pioneering studies (Section 4.3.1).

CHAPTER 7

Contributions of paleoecology to Easter Island's prehistory

Contents

7.1	Discovery and settlement	217
7.2	Climate changes of the last millennia	222
7.3	Spatiotemporal deforestation patterns	225
7.4	Cultural aspects	229
	7.4.1 Polynesian cultigens	229
	7.4.2 Freshwater availability and potential cultural consequences	230

After reviewing in detail the paleoecological studies developed to date on Easter Island, the obvious question is: how does this help understanding island's prehistory? As already stated (Section 3.1.1), paleoecology is unable to provide direct evidence on a number of cultural matters, but it can inform on the relevant ecological, environmental, and human issues—e.g., settlement, vegetation and landscape dynamics, climatic shifts or land use and agricultural practices, among others—that are useful by themselves and also to place cultural developments in the right socioecological scenario. This chapter is an attempt to distillate the most reliable and significant paleoecological information presented in Chapters 4–6, useful to address the main points of the book, namely human settlement, climate change, deforestation patterns, and cultural shifts.

7.1 Discovery and settlement

As we have seen in Section 2.3, the available archaeological evidence preserved on terrestrial settings situates the arrival of Polynesians to Easter Island at some time between 800 and 1200 CE. However, as mentioned in the same section, these dates may be considered minimum ages because terrestrial evidence of human settlement may eventually be removed by natural and/or anthropogenic agents (evidence clearing) or hidden

Paleoecological Research on Easter Island
https://doi.org/10.1016/B978-0-12-822727-5.00007-6

by recent sea-level rise (Stevenson et al., 2000; Flenley and Bahn, 2011; see also Section 3.1.1). In other oceanic islands, it has been demonstrated that lake sediments may preserve evidence for human presence—e.g., discovery without settlement or ephemeral/intermittent settlement—prior to permanent establishment (Rull et al., 2017; De Boer et al., 2019). Could this be the case for Easter Island?

The first well-dated evidence of deforestation and grass meadow expansion was found by Butler and Flenley (2010) in Rano Kao sediments (core KAO2), between 1900 and 1850 cal yr BP, or 50–100 CE. The coincidence of this forest clearing event with a conspicuous charcoal increase (Fig. 6.6) suggested that humans could have been involved. These authors noted that this would be contrary to most archaeological reconstructions and discussed other possibilities for the origin of fire such as spontaneous combustion, a combination of lightning and dry climates, and volcanic eruptions. They argued that lightning is rare on Pacific islands and is always accompanied by heavy rainstorms (Doswell, 2002), which discourages fire, and that there is no evidence of volcanic eruptions by 1900 cal yr BP (González-Ferrán et al., 2004). Butler and Flenley (2010) also emphasized that charcoal concentrations never returned to background levels after the 1900 cal yr BP exacerbation, suggesting continuous disturbance. In addition, they noted that, according to linguistic evidence, human arrival on Easter Island would have occurred by 100 CE (Flenley and Bahn, 2003). The conclusion was that further evidence was needed for a sound interpretation of the available paleoecological evidence, which could not be dismissed. Also in Rano Kao (Core 1), Horrocks et al. (2013) found a high concentration of charcoal fragments (Fig. 6.10), associated to banana (*Musa*) phytoliths, in sediments corresponding to 3680–2680 cal yr BP (1730–910 BCE). According to these authors, this could be interpreted in terms of anthropogenic forest clearing by fire and associated agriculture, but the dates were confusing as they were much older than the expected settlement of Eastern Polynesia (Kirch, 2010; Wilmshurst et al., 2011). Indeed, according to the more accepted chronological framework, the Polynesians could have been colonizing the Central Pacific archipelagos by those times (1730–910 BCE) (Fig. 2.4). Horrocks et al. (2013) did not dismiss the potential existence of dating problems due to contamination (Section 6.2.1) although they maintained the robustness of their age-depth model (Fig. 6.9). None of these Kao records have clear indications of human presence but other records from Raraku and Aroi were more specific in this sense.

A well-dated record of possible human presence on Easter Island earlier than expected was obtained by Cañellas-Boltà et al. (2013) in Rano Raraku (core RAR08). The pollen signal was the same as in Rano Kao—that is, an event of forest decline and grass meadow expansion associated to a charcoal increase—but that time this evidence was combined with a new element: the pollen of *Verbena litoralis* (Figs. 6.16 and 6.17), an anthropochore weed of American origin, associated with a variety of human activities and medical applications, including leprosy (Zizka, 1991), an illness that was especially frequent among the Rapanui during the exploitation of the island as a sheep ranch (Section 1.4.3). This plant—locally known as puringa—is now widespread all over the island, from the coasts to the highest elevations and, until recently, it was thought to have been introduced shortly before 1870 CE (Zizka, 1991). However, the Raraku record showed that the species was present since at least 450 BCE, coinciding with the first deforestation pulse recorded in this catchment (Fig. 6.16). Cañellas-Boltà et al. (2013) discussed whether *V. litoralis* could have reached Easter Island by long-distance dispersal or by human introduction. The coincidence of forest demise, meadow expansion and fire with the presence of this anthropochore weed suggested that humans could have introduced *V. litoralis* in Easter Island by 450 BCE, at least 1500 years before the expected Polynesian settlement. In the beginning, forest disturbance was minimum and *Verbena* was scarce but by 1200 CE forest clearance underwent an abrupt acceleration coinciding with a fire exacerbation and a conspicuous expansion of grasses and *Verbena*. Cañellas-Boltà et al. (2013) suggested that the initial forest disturbance and the introduction of *Verbena* could have been due to the arrival of small human populations, whereas and the further forest declines and the later expansion of grasses and *Verbena* were likely the result of larger populations and the increased human use of the Raraku catchment.

Not only the settlement timing but also the origin of the first settlers was challenged by the Raraku record. Indeed, if the American *Verbena* was carried by humans in 450 BCE, the most likely mechanism was direct transportation by Amerindian people. In other words, Native Americans could have visited Easter Island long before Polynesian arrival (Heyerdahl's hypothesis; Section 2.3.2, Fig. 2.5). Unfortunately, the short sedimentary gap between 500 and 1165 CE (Cañellas-Boltà et al., 2013) prevented to infer whether the assumed Amerindian presence was continuous until Polynesian arrival and whether or not these cultures could have coincided and eventually interacted. However, before 500 CE, the continuity of grasses, *Verbena* and fires suggested maintained but low-intensity disturbance, with forests remaining virtually untouched (Fig. 6.16). In summary, contrarily to the assertion of Flenley and Bahn (2003) that Easter Island was settled only once by Polynesians during the last millennium, the finding of *Verbena* suggested that the island had already been discovered, albeit probably not permanently settled, by Amerindians by 450 BCE. This agreed in part with the ideas of Heyerdahl, except for the date of arrival of Amerindians—400 CE, according to Heyerdahl and Ferdon (1961)—and the eventual violent contact between the two cultures. The whole picture may also be consistent with the idea of Mann et al. (2003) of small scattered and perhaps transient human populations of hunter-gatherers with low-impact practices until 1200 CE when full island disturbance commenced. The novelty introduced by the Raraku record is that Native Americans could have been part of the ephemeral and/or intermittent settlers and that some of the plants they introduced (e.g., the verbena) were adopted by Polynesians and became part of their own culture. It has also been suggested that the arrival of *Verbena* in Easter Island might have been by back-and-forth mechanisms, as it occurred with the sweet potato (Section 2.3.3; Fig. 2.5), which implies that Polynesians arrived first to America and then came back and reached Easter Island (Rull, 2020a). The fact that these initial discoverers and/or settlers have no left terrestrial archaeological evidence might be due to the paucity and ephemerality of their settlements, along with the effect of evidence clearing (Section 3.1.1).

The pollen of another American invader taxa, *Sisyrinchium* (Iridaceae), was found in the sediments of Rano Aroi (cores RA and RAI) by Horrocks et al. (2015) long before Polynesian colonization (Fig. 6.13). The presence of this pollen was consistent in all samples of the RAI core since at least 410–630 CE, which is similar to the settlement chronology formerly proposed by Heyerdahl and Ferdon (1961). Again, these authors maintained the robustness of their age-depth model and considered the possibility of this plant to have been introduced by humans several centuries before the accepted ages for Polynesian colonization. However, these authors were very cautious and argued that "...we report this evidence and note that there is no clear reason to dismiss it, nor there is a clear case to accept it. Specifically, we are unaware of any Rapa Nui archaeological excavations or other evidence that provide conclusive support for prehistoric Amerindian presence" (Horrocks et al., 2015). However, it seems clear that the case of *Sisyrinchium* is compatible with that of *Verbena* and the above comments on the early Amerindian presence on Easter Island. The added value, in this case, is that the RAI core included the time interval hidden by the short sedimentary gap in the Raraku core, thus suggesting the continuity of *Sisyrinchium* during the last 1300–1500 years. This does not mean, however, that Amerindians must have been present on Easter Island during all this time, only that the plant established on the island after its initial introduction, as it occurred with *Verbena*.

Table 7.1 summarizes the events discussed above suggesting early human presence on Easter Island. It seems that novel paleoecological evidence is emerging about the possibility of an early settlement of Easter Island, long before Polynesian arrival, in the form of forest clearing and meadow expansion linked to fire events and accompanied, in some cases, by the introduction of plants of American origin related with human activities. These findings should not be neglected as they are based on robust age–depth models and reliable taxonomic identification. In some cases, the same authors that have recorded these events were too cautious and tried to find alternative explanations because their results contradicted the most accepted archaeological chronologies. However, each approach records what it records. Archaeology records archaeological settlement (Lipo and Hunt, 2016), whereas paleoecology is able to provide information on previous events, with or without archaeological record (e.g., Rull et al., 2017). Paleoecological evidence should not be selected and interpreted according to its agreement or disagreement with current archaeological standards (and vice versa) but should be taken as is, in an objective manner, provided methodological aspects such as those mentioned above are guaranteed. Therefore, testing the hypothesis of early human presence on Easter Island is a suitable paleoecological aim, whose first steps have already been done and should be enhanced and improved. The coincidence of forest clearing events, fire, and the introduction of plants of American origin seem to be a good start, which could be improved with the incorporation of direct evidence of human presence using biomarkers such as DNA and specific fecal lipids, among others (e.g., Bull et al., 2002; D'Anjou et al., 2012; Hofreiter et al., 2012; Rawlence et al., 2014; Parducci et al., 2017). The field of biomarker

Table 7.1 Summary of the events suggesting early human presence on Easter Island before Polynesian settlement.

Site	Age	Events	Introductions	References
Rano Aroi	410–630 CE	Forest decline and fire	*Sisyringium* (America)	Horrocks et al. (2015)
Rano Kao	50–100 CE	Forest decline and fire	N.A.	Butler and Flenley (2010)
Rano Raraku	450 BCE	Forest decine and fire	*Verbena litoralis* (America)	Cañellas-Boltà et al. (2013)
Rano Kao	1730–910 BCE	Forest decline and fire	N.A.	Horrocks et al. (2013)

N.A., not ascertained.

analysis in lake sediments is currently in full swing and it is hoped that future paleoecological studies on Easter Island can take advantage of this.

To conclude this section, it is interesting to reproduce a couple of passages of Mulloy (1974) related to the early occupation of Easter Island and the potential American presence:

> *the earliest reliable radiocarbon date so far discovered is A.D. 690. It establishes the time of construction of one of the many gigantic outdoor altars – and undertaking certainly not the first activity of new immigrants. A community undoubtedly had been developing for many years before this date and it is thus not unreasonable to assume that people were already living here about two thousand years ago.*
>
> *though the stockpile of ideas brought from the earliest homelands might have been rendered more ample and diverse than one would suspect by arrivals from a considerable number of other islands and perhaps the South American continent as well. . . .*

These ideas, which in their time might have seemed too speculative, are beginning to make sense by being able to be tested with paleoecological methods.

7.2 Climate changes of the last millennia

Paleoecological research on Easter Island has also provided local paleoclimatic information using biotic and abiotic proxies independent from pollen and spores, which has two main advantages. First, climatic shifts have been documented in situ, which avoids the need to rely on assumptions derived from regional climatic reconstructions and correlations with distant localities, often governed by disparate climatic systems. Second, the use of physicochemical proxies and microfossils other than pollen and spores to reconstruct past climates allows circumventing circular reasoning and makes it possible to study the response of vegetation and landscape to climate shifts. Mann et al. (2008) emphasized the need for increasing the resolution of the last millennium to clarify the potential influence of eventual droughts on the timing and patterns of the deforestation process (Section 6.1.1). To date, higher-resolution paleoclimatic studies of the last millennia have been published for Rano Raraku and Rano Aroi, whereas Rano Kao is still under study. The main paleoclimatic proxies studied have been sedimentology, C and N isotopes, and diatoms. The following is a synthesis of the main paleoclimatic trends that have been analyzed case-by-case in Chapter 6.

The first general observation is that only changes in the moisture balance have been recorded, which supports the early predictions of Flenley et al. (1991) that temperature shifts on Easter Island would have been minor and, therefore, of little ecological significance. Regarding moisture variations, the most relevant findings have been the occurrence of two centennial-scale droughts separated by a phase of wet climates (Fig. 7.1). The first drought, called the MCA drought, was recorded in Rano Raraku between 500 and 1170 CE, when the lake dried out and remained in this condition for more than six centuries (Cañellas-Boltà et al., 2013). The same drought was recorded at Rano Aroi between 600 and 1100 CE using isotopes (Rull et al., 2015). It is interesting to note that the MCA drought was coeval with the Classic Maya Collapse (c.900 CE), which was attributed to a series of prolonged droughts that led to the final demise of this civilization (Haug et al., 2003). An ensuing moisture increase was recorded between about 1170 and 1570 CE at Rano Raraku by a significant water-level rising, as deduced from a conspicuous shift in the diatom flora. The same wettening

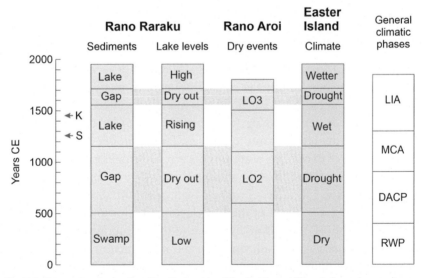

Fig. 7.1 Reconstructed paleoclimatic trends of the last two millennia based on sedimentology, physicochemical proxies and biological proxies other than pollen and spores in Rano Raraku and Rano Aroi (cores RAR08 and ARO08-02, respectively). Raw data from Cañellas-Boltà et al. (2013) and Rull et al. (2015) discussed in Chapter 6. Droughts are highlighted by *gray* bands. *Blue* arrows are regional volcanic eruptions occurred during Easter Island's prehistory (K, Kauwe eruption; S, Samalas eruption) (Margalef et al., 2018). DACP, Dark Ages Cold Period; LIA, Little Ice Age; MCA, Medieval Climate Anomaly; RWP, Roman Warm Period.

was observed in Rano Aroi after isotope analysis. This wet phase could have included the questioned 1300-event of Nunn (2000, 2007) and Nunn and Britton (2001), characterized by uncommonly heavy precipitation, as a consequence of an increase of the El Niño frequency (Section 3.2).

The second drought, the LIA drought, occurred between about 1570 and 1720 CE, when Lake Raraku dried out again (Fig. 7.1). In Rano Aroi, this drought was more difficult to detect, probably due to the masking effect of human disturbance since 1520 CE onward. The LIA drought coincided with the so-called "17th century crisis" in tropical Asia, characterized by droughts, famines, and large-scale economical and political disruption (Grove and Adamson, 2018). This crisis has been related to changes in the variability of the Indian Ocean Dipole (IOD) climatic system affecting ENSO variability in the Pacific Ocean (Abram et al., 2020). Before the intense MCA and LIA droughts, Easter Island climates were drier than at present (but not arid) in both Raraku and Aroi (Cañellas-Boltà et al., 2013; Rull et al., 2015). After the LIA drought, Lake Raraku attained its present-day levels, which suggested the returning of wetter climates. The potential climatic mechanism causing the droughts have already been discussed in Section 6.1.1 and can be summarized in a southward shift of the humid subtropical storm track and the emplacement of the dry South Pacific Anticyclone (Fig. 1.5) over Easter Island, likely forced by ENSO fluctuations (Mann et al., 2008; Sáez et al., 2009).

In a recent paper, Margalef et al. (2018) suggested that some major volcanic eruptions occurred in Pacific islands might have affected regional climatic trends by lowering average temperatures and/or by affecting ENSO variability and, therefore, moisture patterns. These authors emphasized the potential effect of two major eruptions occurred during Easter Island's prehistory, in the Samalas (1257 CE) and the Kuwae (1453 CE) volcanoes, the first situated in the Indonesian Lombok Island, east of Bali, and the second in the Vanuatu Islands (Fig. 1 of the Introduction). Unfortunately, according to the same authors, there is no evidence of these eruptions in the Easter Island's sedimentary cores studied to date and it is difficult to find evidence of the potential influence of regional volcanism on Easter Island's climates. The dates of the mentioned volcanic events are incorporated into Fig. 7.1 in the hope that future research will be able to test their eventual impact on climatic trends of the island. Noteworthy, both eruptions occurred during the wet phase of rising lake levels between the MCA and LIA droughts.

There is still much work to do for a sound understanding of recent paleoclimatology of Easter Island, especially in terms of resolution and spatial patterns. At present, the Rano Kao KAO08-03 core (Fig. 6.19) is being studied for high-resolution paleoclimatic reconstruction using the isotopic composition of plant leaf waxes as precipitation proxies, but others organic biomarkers are also available for similar purposes (Eglington and Eglington, 2008; Castañeda and Schouten, 2011). These techniques should also be applied, if possible, to other cores already obtained and other news that can be retrieved in further fieldwork.

7.3 Spatiotemporal deforestation patterns

It is not unusual to believe that Easter Island's deforestation was a single event across the island with synchronous and well-defined beginning and end (e.g., Flenley et al., 1991; Mann et al., 2003; Hunt, 2006; Mieth and Bork, 2010), although the specific dates may vary according to authors (Section 2.4.4). However, paleoecological research has shown that forest clearance did not occur at the same time and at the same rates in Rano Aroi, Rano Kao, and Rano Raraku catchments (Rull, 2016b, 2020a). The purpose of this section is to strengthen this point and also to explore the possible influence of climate shifts and human activities on deforestation patterns. The discussion will prioritize paleoecological reconstructions from continuous (gap-free) and chronologically coherent (no age inversions) lake and swamp cores like those retrieved during the revival phase (Chapter 6). Section 7.1 already summarized deforestation events recorded before Polynesian settlement (800–1200 CE). In this section, emphasis will be placed on the last millennium, which is also the time interval that was usually hidden by depositional gaps or disturbed by sediment mixing in former Easter Island records. Figs. 7.2 and 7.3 clearly show that deforestation was heterogeneous in time and space.

The first deforestation pulse during or after Polynesian settlement was recorded in Rano Kao at about 1050 CE (Seco et al., 2019). As stated above (Section 6.2.2) humans were already present around the lake but in very low numbers, likely insufficient to cause the magnitude of the deforestation observed. This deforestation pulse occurred during the MCA drought that dried out Lake Raraku, which suggests a potential influence of climate. Rano Aroi was unaffected likely by the maintenance of enough moisture for forest development thanks to the elevational precipitation gradient, which today is close to 200 mm/100 m elevation (Puleston et al., 2017). This first local deforestation event in Rano Kao occurred during the phase of low-intensity forest disturbance proposed by Mieth and Bork (2015), and more than a century before the dates proposed by the defenders

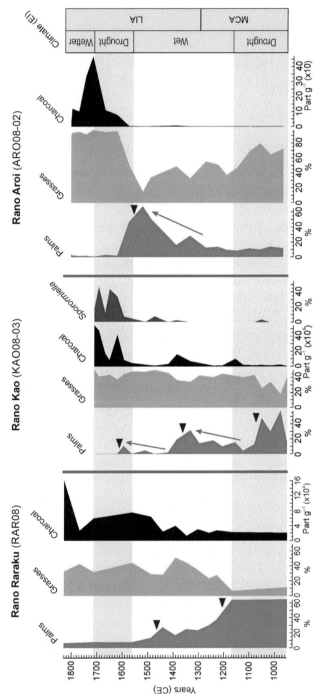

Fig. 7.2 Synthetic diagrams of Rano Raraku (Cañellas-Boltà et al., 2013), Rano Kao (Seco et al., 2019), and Rano Aroi (Rull et al., 2015) for the last millennium (950–1850 CE). Droughts are depicted as gray bands. *Red triangles indicate deforestation pulses, and forest regeneration trends are indicated by blue arrows.* EI, Easter Island. *(Redrawn and modified from Rull, V., 2020. The deforestation of easter Island. Biol. Rev. 95, 121–141.)*

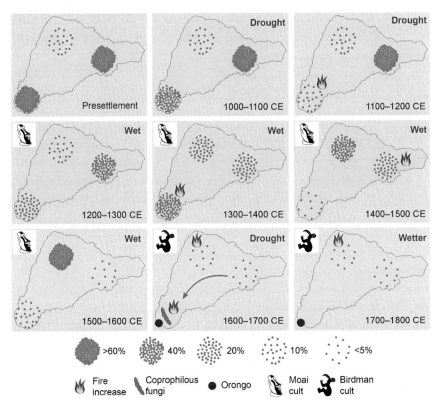

Fig. 7.3 Spatiotemporal deforestation patterns of Rano Aroi, Rano Kao, and Rano Raraku areas and their potential drivers after Polynesian settlement, using the information in Fig. 7.2. Forest cover was estimated by palm pollen percentage. The *blue* arrow at 1600–1700 CE indicates the relocation of the cultural core of the Rapanui society from Rano Raraku to Rano Kao. *(Redrawn and modified from Rull, V., 2020. The deforestation of easter Island. Biol. Rev. 95, 121–141.)*

of late colonization, in 1200 CE or after (Hunt and Lipo, 2006; Wilmshurst et al., 2011). This phase could be one more manifestation of human settlement without extensive archaeological expression and/or further evidence clearing.

The next deforestation episode occurred in Raraku at around 1200 CE, just after the MCA drought, under sustained, although moderate, fire pressure (Cañellas-Boltà et al., 2013). The occurrence of wetter climates more favorable for forest growth suggests that this deforestation pulse was of anthropogenic origin. Forest retreat was continuous despite fires did not increase significantly, which suggests the existence of positive feedbacks amplifying forest responses to sustained fire incidence. At the same time, the Kao forests were recovering, likely due to favorable climates and the absence of human pressure, as indicated by the lack of fires and coprophilous fungi (Seco et al., 2019). However, this forest regeneration was interrupted by another deforestation event (1350 CE), likely anthropogenic, coinciding with a charcoal peak and the return of *Sporormiella* spores (Seco et al., 2019). Rano Aroi continued to be devoid of humans, and the significant forest expansion recorded between approximately 1300 and 1500 CE was likely due to the occurrence of wetter climates (Rull et al., 2015).

Rano Raraku forests disappeared after a final deforestation event occurred by 1450 CE, coinciding with significant fire exacerbation supporting human deforestation (Cañellas-Boltà et al., 2013). At the same time, human pressure declined in Rano Kao, and its forests, which had been almost removed during the 1350 CE deforestation pulse, experienced a new regeneration trend although less intense than the former (Seco et al., 2019). At the same time, Rano Aroi was truly forested for the first time, and its forests became the densest and most extensive on the island due to the almost complete deforestation of Raraku and Kao catchments. This forest densification could likely be due to wetter climates and the absence of human disturbance in Rano Aroi. However, the situation changed by 1570 CE, when a sudden deforestation event completely removed these forests in about a century (Rull et al., 2015). This coincided with intense fire exacerbation, suggesting anthropogenic causes; however, the LIA drought would have acted synergistically by favoring forest flammability and preventing regeneration. The last deforestation event on the island irreversibly removed the Kao forests by 1600 CE, coinciding with the largest human occupation of this crater, as indicated by the significant increase in fires and *Sporormiella* spores. This has been associated with the first activities around the ceremonial village of Orongo, situated at the SW crest of the Kao crater (Robinson and Stevenson, 2017).

Taken globally, the intensification of deforestation between approximately 1200 CE and the total disappearance of forests by around 1600 CE coincides with the phase of increased forest clearing formerly proposed by Mann et al. (2003) and Mieth and Bork (2015), after a phase of low-intensity ecological impact (Section 2.2). However, the detailed reconstruction shown here, based on continuous sedimentary records from the three basins suitable for palaeoecological study, significantly increases the spatiotemporal resolution and provides much more detail on the patterns, processes, and possible causes of the different forest removal and regeneration events. In addition, forest regeneration processes, as formerly proposed by Mieth and Bork (2010) using edaphological studies, are strongly supported by palynological records. Further efforts should emphasize the study of proxies for human presence such as *Sporormiella* and specific molecular biomarkers. The use of paleoclimatic biomarkers and an increase in temporal resolution is also recommended.

7.4 Cultural aspects

Paleoecology has also provided direct information on cultural aspects of the ancient Rapanui society, mainly on cultivated plants. In other cases, paleoecological information has been useful to analyze new cultural scenarios emerging from previously unnoticed events. The following sections summarize these aspects.

7.4.1 Polynesian cultigens

Microfossil analysis of lake sediments and drylands across Easter Island has provided direct evidence of plants cultivated by the ancient Rapanui, as a sample of Polynesian cultivation practices (Table 7.2). As discussed in Section 2.4.5, cultivation was mainly performed on protected manavai but paleoecology demonstrated that lakeshore terraces were also actively cultivated, possibly using local irrigation techniques, mainly in Rano Kao and Rano Raraku (Horrocks et al., 2012a,b, 2013; Sherwood et al., 2019), where freshwater was easily accessible. The available evidence indicates that cultivated species were widespread across the island with no manifest crop specialization between drylands and lake terraces. However, it is noteworthy that some species (yam, bottle gourd, and sweet potato) were not cultivated in Rano Aroi, which suggests that these cultigens were probably restricted to lowlands. Unfortunately, a chronology of ancient Rapanui cultivation cannot still be established because of the already discussed dating difficulties. This would be a very interesting target to be pursued. The study of pollen, phytoliths, and starch of cultivated plants in continuous and chronologically consistent cores used to unravel deforestation patterns (Section 7.3) could be a good start, which might be complemented with new coring campaigns along lakeshores.

Table 7.2 Plants cultivated by the ancient Rapanui as recorded from pollen, phytoliths and starch in lake/swamp sediments and in dryland soils.

Taxa	Local name	Site	Proxy	Reference
Dioscorea alata	Yam	Dryland soils, lake sediments (Kao)	Starch	Horrocks and Wozniak (2008) and Horrocks et al. (2012b)
Lagenaria saceraria	Ipu kaha (bottle gourd)	Dryland soils, lake sediments (Kao, Raraku)	Pollen	Horrocks and Wozniak (2008) and Horrocks et al. 2012a)
Ipomoea batatas	Kumara (sweet potato)	Dryland soils, lake sediments (Kao, Raraku)	Pollen, starch	Horrocks and Wozniak (2008) and Horrocks et al. (2012a,b, 2016)
Colocasia esculenta	Taro	Dryland soils, lake/swamp sediments (Aroi, Kao, Raraku)	Pollen, starch	Horrocks and Wozniak (2008) and Horrocks et al. (2012a,b, 2015, 2016)
Broussonetia papyrifera	Mahute (paper mulberry)	Dryland soils, lake/swamp sediments (Aroi, Kao)	Phytoliths	Horrocks et al. (2012b, 2016)
Musa sp.	Maika (banana)	Dryland soils, lake/swamp sediments (Aroi, Kao, Raraku)	Phytoliths	Horrocks et al. (2012a,b, 2015, 2016)

7.4.2 Freshwater availability and potential cultural consequences

As we have seen in Section 1.2, permanent freshwater sources are scanty on Easter Island due to the high permeability of its porous volcanic rocks (Herrera and Custodio, 2008). Rano Kao and Rano Raraku are fed solely by rainfall and groundwater is accessible in Rano Aroi and on the coasts, where wells for human use are common today (Fig. 1.7). During the MCA and the LIA droughts, Rano Raraku was totally dry, and freshwater availability was critical for human life. Polynesians arrived during the MCA drought when Lake Raraku was dry but its surroundings were still forested.

At that time, the only permanent freshwater sources were Lake Kao and the Aroi swamp. Not surprisingly, the Kao forests were the first to be disturbed, probably by small populations arriving in the island (Fig. 7.3). During the wet phase following the MCA drought, Lake Raraku replenished and the forests around started to be cleared. This occurred by 1200–1300 CE, coinciding with the onset of monumental architecture and the moai cult (Fig. 2.21). The continued deforestation of Rano Kao and Rano Raraku left their forests close to the disappearance by 1500–1600, whereas Rano Aroi, still free from human influence, experienced forest densification likely owing to wet climates.

In these conditions arrived the LIA drought and the second drying out of Rano Raraku, which was already deforested since 1450 CE (Cañellas-Boltà et al., 2013). The situation was more critical than during the MCA drought, as the Rapanui population was more numerous and Rano Raraku, the center of the moai industry and the core of the Rapanui culture, was likely a badland devoid of freshwater and forests (Rull, 2020b). Again, the only permanent freshwater bodies were Lake Kao and the Aroi swamp (Rull, 2016a), which began to be fully exploited and were totally deforested by 1600 CE (Fig. 7.3). Rano Aroi was probably abandoned after deforestation but human activities continued in Rano Kao, likely leading to the emergence of the Birdman cult (Fig. 2.21) and the foundation of its cultural center: the ceremonial village of Orongo (Robinson and Stevenson, 2017) (Fig. 2.14). Additional evidence of human activity is the cultivation of Kao lake margins just below Orongo (Fig. 7.4). Therefore, anthropogenic deforestation of Rano Raraku (1450 CE) followed by the incoming of the LIA drought (1570 CE) might have been involved in the geographical shift

Fig. 7.4 Panoramic view of Rano Kao showing the position of the ceremonial village of Orongo, on top of the SW crater crest, and the location of formerly cultivated terraces *(yellow line)*, as documented by paleoecological records (Horrocks et al., 2012b).

of the Rapanui cultural center. This supports the former idea of McCall (1993), according to whom, the LIA should not be viewed as a decisive deforestation driver but as a new factor to be considered in the explanation of further social and cultural developments (Section 3.2).

The causes of the shift from the moai cult to the Birdman cult remains controversial but several reasons may be suggested from the above observations. The first is, obviously, the search for freshwater but this only justifies the geographical displacement of the cultural core of the Rapanui society. Another argument is that the Kao rocks are not suitable for moai carving. Indeed, the Kao crater is made of hard basalt and is one of the quarries for obtaining the tools (toki) to carve the softer Raraku tuff (Gioncada et al., 2010). As mentioned before (Section 1.5), only 13 of the c.1000 moai known are made of basalt (Van Tilburg, 1994). Robinson and Stevenson (2017) suggested that the shift from the moai cut to the Birdman cult was a territorial restructuring in response to soil nutrient depletion in interior lands, probably due to deforestation and a long period of dryness (Section 2.4.3), which is in agreement with paleoecological observations. The combination of these climatic and anthropogenic disturbing factors, however, did not lead to the collapse of the Rapanui society, which was able to maintain a healthy society on the basis of new agricultural developments (Mulrooney, 2013; Mieth and Bork, 2015, 2017; Stevenson et al., 2015; Jarman et al., 2017; Wozniak, 2017) (see discussion in Section 2.4.5). It has also been proposed that the shift from the moai cult to the Birdman cult was a change from a rigid hierarchical and dynasty-based society to a more dynamic and democratic sociopolitical organization, which represented a strategy for a better adaptation to the new environmental context. In this way, the Rapanui neither collapsed nor disappeared but remained in their home despite the intense environmental pressure, showing exceptional resilience (Rull, 2016b). As formerly pointed out by Bahn and Flenley (1992), after all, the prehistoric Rapanui achieved sustainable stability after deforestation (Section 4.4).

Brosnan et al. (2018) have a different view on how the Rapanui dealt with the freshwater crisis of the LIA drought. These authors considered that fresh/brackish coastal seeps were the major sources of freshwater for the survival of the ancient Rapanui society, as other potential sources (permanent lakes, springs, ephemeral streams, and pools) would have been insufficient to support a population of thousands of individuals. They considered lakes and swamps too remote (Aroi) or difficult to access (Kao) to be routine freshwater sources. In addition, Brosnan et al. (2018) highlighted the lack of human habitation remains in the shores of Lake Kao. Temporary streams were considered too ephemeral and the small stone reservoirs called taheta (Section 1.5) too small and susceptible to evaporation. Therefore, these authors suggested that the main freshwater sources were coastal seeps fed by the groundwater system, where freshwater accumulates on top, and salinity increases with depth due to the diffusion of seawater from below, which creates a vertical density gradient (Fig. 1.7). Present-day wells contain small amounts of freshwater, less than a meter deep, above brackish and marine groundwater (Brosnan et al., 2018). The Rapanui did not have the technology to drill deep wells in volcanic rocks. To capture fresh and brackish waters from coastal seeps, they used pits excavated parallel to the shoreline. The remains of these structures, called puna, have been found at several sites along the island's coasts, which were the preferred sites for the Rapanui to live. Therefore, fresh/brackish water sources would have been frequent, widespread, and close to the populated sites (Brosnan et al., 2018). According to DiNapoli et al. (2018), the ahu, in addition to their ritual meaning, would have a signaling function to indicate the locations of such coastal seeps. All waters found today in coastal seeps are brackish (c.4–28 g of marine salt/L, compared to 1 g/L or less for freshwater and 35 g/L for seawater), which led Brosnan et al. (2018) to suggest that the Rapanui drank brackish water, a fact that, according to these authors, has been well documented historically.

But the view of Brosnan et al. (2018) is not incompatible with that of Lake Kao as a freshwater source. First, it is true that the inner Rano Kao walls are high and steep (Fig. 7.4) but not impracticable, at all, as it is possible to easily descend to the lake and come back by foot in approximately one hour or less. Modern paleoecologists know this well, and we use this method during sediment-coring campaigns with all the equipment and necessary provisions. The ancient Rapanui demonstrated an outstanding transporting capacity by moving the moai from the Rano Raraku quarry to all parts of the island, including elevations above 200 m (Fig. 1.1). Transporting water across the Rano Kao walls is a much easier task that could be conducted on a daily basis. Given the absence of conspicuous water-transporting infrastructures (Hunt and Lipo, 2011; Puleston et al., 2017), freshwater might have been carried in easily transportable containers such as bottle gourd (*Lagenaria siceraria*) fruits, which were intensively cultivated for such purpose across Polynesia, including on Easter Island (Clarke et al., 2006; Green, 2000b; Horrocks and Wozniak, 2008; Horrocks et al., 2012a). Easter Islanders could have easily transported water from Lake Kao in this way. Indeed, practitioners of the Birdman cult were required to descend to the sea, swim to the Motu Nui islet (Fig. 2.11), situated 1.5 km away from the coast, and return to Orongo through the outer Rano Kao cliffs, which were remarkably higher and more difficult to climb than the inner walls of this crater. Obtaining freshwater from the lake is a much easier task that could be routinely performed by common people. It seems totally unreasonable that Rapanui people would have refused to routinely obtain water from Rano Kao during a prolonged drought as the centennial-scale LIA drought, just to avoid a barely 1-hour trip to the lake shores.

In addition, evidence of precontact agriculture along the Rano Kao shores immediately below Orongo (Fig. 7.4) suggests a connection with this relict village, which has the shorter and easiest access to the lake. There is no evidence of human dwellings on the lake shores, and it is reasonable to assume that the agricultural products were transported to Orongo for human consumption and, eventually, distribution to other places. The same would be true for lake freshwater, whose transportation could have been performed in the same way and at the same time. Transporting agricultural products but no freshwater from the Rano Kao shores seems absurd. Given the already mentioned outstanding transporting capabilities of Rapanui people, distributing agricultural products and water from Orongo to other places would have been very easy. The main strength of the coastal groundwater hypothesis (Brosnan et al., 2018) is that water sources are numerous, widespread, and close to habitation sites. However, there are two main drawbacks in considering this possibility as the only way of obtaining freshwater during the LIA drought. First, this hypothesis was erected after a detailed study of only the eastern half of the island, where access is easier. Given the present state of knowledge, this situation cannot be extrapolated to the western sector, which included the center of the Rapanui culture during the LIA drought, where the physiography is very different and coastal seeps have not been reported. Second, all present-day coastal seeps identified in the eastern sector produce brackish water rather than freshwater (Brosnan et al., 2018). If this was the main water source for the ancient Rapanui during the LIA drought, they must have survived for approximately six generations (150 years; 1570–1720 CE) with only brackish water for drinking and agriculture, which would have been challenging, if not hard to believe, and does not seem to be the most efficient solution. Finally, as rain is the only freshwater source for the groundwater system, such a supply should have been drastically reduced during droughts, suggesting that the salinity of coastal seeps could have been higher than it is today. Therefore, coastal seeps could have been used by the ancient Rapanui, but other truly freshwater sources likely would have been needed to maintain this society in good shape.

In summary, the hypotheses discussed here have advantages and drawbacks, but with the available evidence, neither can be rejected. Therefore, there is no reason to exclude either of these hypotheses to explain freshwater availability during the LIA drought. From a human perspective, it seems reasonable to take advantage of any freshwater sources available during a centennial-scale drought such as that during the LIA. It should be noted that the hypotheses examined here are testable rather than speculative. A simple test would be a long-term experiment involving two human populations and their descendants, one entitled to use all natural freshwater sources available on the island and the other relying only on brackish coastal seeps for a century and a half. It would be interesting to see how many participants would voluntarily enroll in the second group.

CHAPTER 8

From human determinism to environmental, ecological, and social complexity

Contents

8.1	The EHLFS approach	237
	8.1.1 General considerations	237
	8.1.2 Multiple working hypotheses and strong inference	242
	8.1.3 The EHLFS approach in Easter Island	243

At the end of Chapter 2 (Section 2.5), it was asked whether climate change could have had some influence on Easter Island's prehistoric ecological and cultural transformations or, on the contrary, all that happened was caused by human activities (human determinism). In the same section, it was also stated that the approach of the book was free of any a priori determinisms, including environmental determinism. Now we have the necessary elements to start analyzing the climatic, ecological, and cultural developments of Easter Island's prehistory under a holistic perspective. For this to be achieved, we will use the EHLFS (Environmental-Human-Landscape Feedbacks and Synergies) approach (Rull, 2018), which is a generalization of the CLAFS (Climatic-Landscape-Anthropogenic Feedbacks and Synergies) approach (Rull et al., 2018).

8.1 The EHLFS approach

8.1.1 General considerations

EHLFS could be viewed as a functional system formed by three basic components or subsystems, namely the environment (E), the humans (H), and the landscape (L), as well as their corresponding interactions, as expressed in all the potential feedbacks (F) and synergies (S) within and among them (Fig. 8.1). The three basic subsystems (E, H, and L) are composed of many elements, some of which are particularly relevant to define and characterize

Paleoecological Research on Easter Island
https://doi.org/10.1016/B978-0-12-822727-5.00008-8

EHLFS
Holistic approach

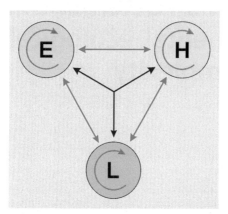

Fig. 8.1 Schematic view of the EHLFS holistic approach. E, Environment; H, Humans; L, Landscape. Feedbacks are represented by *blue* arrows and synergies by *red* arrows. *Gray arrows represent internal subsystem feedbacks. (Redrawn from Rull, V. 2018. Strong Fuzzy EHLFS: a general conceptual framework to address past records of environmental, ecological and cultural change. Quaternary 1, 10.)*

their interactions. Major environmental elements are climate change, geological patterns, or environmental hazards (volcanism and earthquakes), as well as all the associated astronomical, atmospheric, oceanic, and lithospheric drivers and processes. The human component is represented mainly by activities related to land use, occupation and transformation, and the related processes, notably the exploitation of natural resources, demographic changes, technological improvements, migratory patterns, war and other societal conflicts, and communication networks. Rather than a merely descriptive unit, the landscape is treated here as an ecological component, that is, a functional entity formed by the assembly of the different ecosystems that live together in a given region and interact with each other. In terrestrial ecosystems, vegetation is a major landscape feature and its dynamics over time and space is commonly used as a proxy for general landscape dynamics.

Internal feedbacks within E, H, and L influence the output of each of these subsystems and, therefore, the nature of the interactions among them (external interactions). A relevant feature of internal and external feedbacks is the occurrence of amplification mechanisms that can lead to unexpected outputs as a result of nonlinear responses (Overpeck, 1996; Overpeck et al., 2003; Bradley et al., 2003; Williams and Jackson, 2007). For example, a shift to more arid climates (E) can cause changes from forested to more open landscapes (L), which may increase evaporation and enhance local climatic aridification thus amplifying the initial climatic signal and triggering positive feedbacks that can eventually lead to desertification. In this case, climate is the initial forcing factor but landscape features also influence climatic trends at local and regional scales. Similar amplification processes can occur between H and L in the case of human deforestation by fire, which enhances vegetation flammability and exacerbates fire proliferation. The concurrence of climatic dryness and human deforestation is an example of synergy, in this case between E and H acting on L, whose devastating potential is notably enhanced by the coupling of multiple feedbacks among the three subsystems. In this case, landscape changes (L) affect both climate (E) and humans (H), as deforestation patterns may influence, for example, human settlement, demography, and land use.

The influence of human activities on environmental and landscape features has increased throughout the Quaternary. At the beginning of the Pleistocene, with the *Australopithecus* and the first *Homo* species, the influence was absent or negligible. In such pristine condition, the H element was virtually absent from the system, which was composed only by the E and L subsystems and their internal and external feedbacks. Synergies were still absent and the system was an ELF (Fig. 8.2). With time, human influence increased and the system became complete. Human influence became more decisive in the Middle Holocene, after the worldwide establishment and expansion of the agriculture and the incoming of sedentary societies, also known as neolithization (Roberts, 2014). Since then, the Earth has experienced a profound anthropogenic transformation, mainly on the L subsystem. During the last centuries, the Industrial Revolution and the Great Acceleration have resulted in a new state in which humans have dramatically increased their impact until the point of affecting global patterns and processes, including the functioning of the whole Earth System (Zalasiewicz et al., 2011).

Simple and dual deterministic approaches may be viewed as a priori simplifications of the EHLFS system. For example, the so-called environmental determinism considers that environmental changes are the main drivers of ecological turnover, landscape degradation and/or cultural disruption. This model has been used to explain the collapse of past societies and civilizations—for example, the Akkadian Empire in the Near East or the Classic Maya empire in Mesoamerica—through the Holocene due to the occurrence of unexpected climatic events, notably droughts (DeMenocal, 2001). Environmental determinism has been criticized arguing that it does

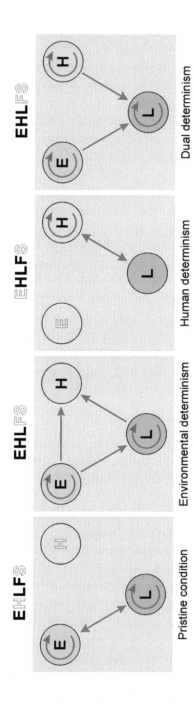

Fig. 8.2 The pristine condition and the more frequent types of determinism, under the general EHLFS framework. *(Redrawn from Rull, V. 2018. Strong Fuzzy EHLFS: a general conceptual framework to address past records of environmental, ecological and cultural change. Quaternary 1, 10.)*

not consider the complexity of human societies and their relationships with the ecological systems and the environment. However, environmental determinism survives because most critiques are based on logical reasoning and the evidence needed to falsify it (sensu Popper, 1959) is still unavailable. This is a typical case of determinism due to different disciplinary approaches, as the dominance of environmental changes on human affairs is usually defended by a number of palaeoclimatologists, whereas criticism comes mainly from humanities and social sciences. Within the framework of this study, the environmental determinism may be considered an EHLS system, where the environment (E) acts directly or indirectly—via landscape (L) degradation—on human societies (H), with no feedbacks between humans and the other subsystems (Fig. 8.2). The incorporation of sound evidence from the human subsystem is needed to add these feedbacks, as well as the internal H feedbacks, and to properly analyze the problem from a more general perspective.

Human determinism attributes ecological and cultural changes to anthropogenic causes and neglects the potential effects of eventual environmental shifts. A classic example is the human settlement of remote oceanic islands and the ensuing landscape and societal changes. For example, in the remote archipelagos of the Pacific Ocean, Mid-to-Late Holocene human colonization resulted in catastrophic landscape shifts and biodiversity declines mainly due to deforestation, overhunting and the introduction of alien predators and competitors (Prebble and Wilmshurst, 2009). Anthropogenic transformations have been so intense and extended that the potential effects of environmental changes on the landscape seem to have been negligible. In this case, human action is clearly visible in the palaeoecological record and cannot be ignored, but a potential role for environmental changes such as the MCA or the LIA, among others, is rarely falsified. A number of researchers believe that climatic forcing has been negligible but this view has not yet been properly tested. Others contend that climate would have had a role but the magnitude of anthropogenic impact obscures the palaeoecological record of such influence. In the present state of knowledge, human determinism seems to be the winning option, by far. In this case, the system is an HLF, with only two subsystems (H and L) and their corresponding feedbacks but no synergies. Internal landscape feedbacks are negligible as landscape patterns and processes are considered to be fully controlled by human activities.

Dual controversies are also frequent. A typical case is the origin of the high tropical diversity, which has been attributed to either pre-Quaternary tectonic and palaeogeographic changes by some authors, or Quaternary climatic shifts by others (Rull, 2011). The debate between the potential role of either environmental shifts or human activities on the landscape and its ecosystems is also a hot topic. For example, some authors contend that the Mediterranean biome has been originated primarily by the action of a progressive aridification process starting in the Middle Holocene, while others believe that human activities, notably fire, have been more decisive (Jalut et al., 2009; Vannière et al., 2011). An example that is now under vivid discussion is the natural or anthropogenic nature of the hyperdiverse Amazonian rainforests. Such forests have been traditionally considered among the most pristine ecosystems on Earth but recent studies have suggested that pre-Columbian indigenous people were more numerous than previously thought and are largely responsible for the present composition of Amazonian forests (Lewis et al., 2017). This view has been challenged by those who consider the evidence insufficient to support such a proposal and who rely on natural drivers to explain the current Amazonian biodiversity patterns (McMichael et al., 2017). Often, selective evidence is used to support either proposal but, to date, neither of them has been able to be falsified. As in the case of environmental determinism, this type of controversy falls within an EHL framework, but this time the focus is on the landscape (L), rather than on humans (H), and the feedbacks and synergies between environment (E) and humans are rarely considered.

Downgrading the EHLFS system to more simple and deterministic EHLS, HLF, or EHL versions reduce the interpretive potential and may bias the output toward subjective interests. The use of the full EHLFS framework provides a more general scope to account for the whole body of available evidence. If eventually, after the use of the EHLFS approach, a simpler ELF, EHL, or HLF framework emerges, this would be considered a sound output favoring determinism. But if a given simplification is imposed from the beginning as an axiomatic premise, the output can only be subjective. Two complementary approaches to enhance the power of the EHLFS system are the multiple working hypotheses (MWH) (Chamberlin, 1965) and the Strong Inference (SI) (Platt, 1964) methods, which are briefly explained below.

8.1.2 Multiple working hypotheses and strong inference

To exploit its full potential, the EHLFS approach should be combined with the MWH and the SI frameworks (Rull, 2018). According to Chamberlin (1965), when we develop a theory that seems satisfactory to explain a given phenomenon, there is the danger of remaining too attached to it with a sort of parental affection that may lead us to unconsciously select and magnify the supporting evidence, neglecting the empirical data that could contradict our intellectual child. Sooner or later, this favorite theory becomes a ruling theory, that is, a theory that controls and directs further research, no matter if it is built on sound evidence or is a premature explanation based on insufficient empirical data. Eventually, a ruling theory may turn into a paradigm around which research is organized, and a yes-and-no debate between defenders and detractors establishes, thus blocking eventual progress toward alternative explanations. If eventually, an alternative theory emerges, the controversy may turn into a dual debate between the supporters of either one or the other ruling theory, which is still a deterministic research framework that ignores other possibilities. In either case, a debate exists that may lead to subjective personal or partisan discussions where the objective is to be right and the opponents wrong, rather than to find the better explanation for the observed phenomenon (Harrison, 2011). Such an endless loop can only be broken by a change of mindset toward a more open-minded framework, as is the MWH, where any possible explanation is explored and every testable hypothesis is developed. This approach promotes thoroughness, suggests lines of inquiry that might otherwise be overlooked and develops the habit of parallel and complex thought. However, finding positive support for either one or more hypotheses is not enough to refute the others as, according to the SI principle, Popperian falsification should be explicit and must include all the alternative hypotheses possible (Platt, 1964). In other words, in the MWP-SI framework, hypothesis testing requires not only positive evidence for a given proposal but also explicit falsification of the competing explanations. It should also be noted that hypotheses must not necessarily be considered always contradictory, as a number of them could be complementary and may eventually be united in a single, more general, explanation. This is usually the case of too simple and excluding hypotheses, which often generate fake dual determinisms but, with time, they end by being incorporated into a more general explanatory framework.

8.1.3 The EHLFS approach in Easter Island

On Easter Island, none of the existing proposals about settlement and collapse has explicitly been falsified and, as a consequence, they have survived until today as ruling theories. Examples are the assumedly unique colonization of Easter Island from Polynesia (Section 2.3) or the ecocidal theory for the cultural collapse after island-wide deforestation (Section 2.4). Dual determinism has also persisted for a long time regarding the deforestation causes and the demise of the ancient Rapanui culture. Competing hypotheses have been the ecocidal and the resilient/genocidal theories, both human-deterministic (Fig. 8.3). Within the EHLFS approach, the first is a pure HLF system (without

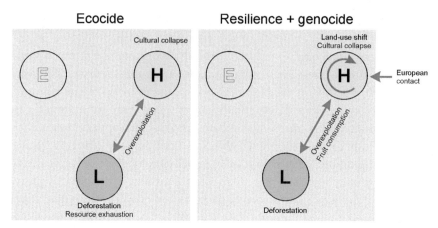

Fig. 8.3 Human-deterministic approaches to Easter Island's deforestation and cultural collapse under the EHLFS framework. Note that the environment is not considered in any of them. *(Redrawn from Rull, V. 2018. Strong Fuzzy EHLFS: a general conceptual framework to address past records of environmental, ecological and cultural change. Quaternary 1, 10.)*

synergies), whereas the second introduces a new element external to the HLF system, as is alien cultural pressure causing a genocide. As in the case of ruling theories, both have supporting evidence but none has still been falsified and, therefore, there is no compelling reason to dismiss any of them. This does not mean that any of them could be acceptable, only that there is no empirical proof for any of them to be false, as required by the strong inference principle (Platt, 1964). But falsification of one of these theories is not enough to accept its competing explanation, as many other possibilities may exist that should be properly tested, according to the MWH framework.

For example, to accept either the ecocidal or the resilient/genocidal hypothesis, it should be demonstrated that climates remained constant or that eventual climate changes had no influence on deforestation and cultural change. Paleoecological evidence reviewed in this book favors the occurrence of climate changes during Easter Island's prehistory, which opens a new perspective able to generate a set of new potential explanations, under a complete EHLFS system. To proceed with this holistic analysis, the first step is to define the specific problem under study, which in this case is the potential consequences of climate changes on deforestation and cultural shifts during Easter Island's prehistory. The second step is to place the archaeological, anthropological, and historical information discussed in Chapter 2 in a paleoclimatic and paleoecological context, using all the evidence gathered in Chapters 4–6 and summarized in Chapter 7 (Fig. 8.4).

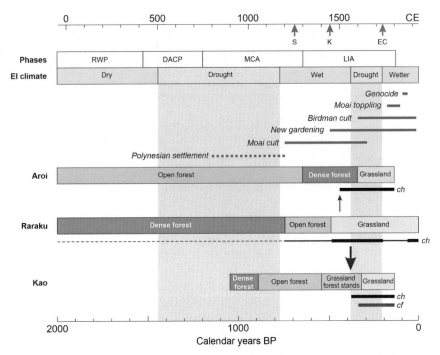

Fig. 8.4 Paleoclimatic and paleoecological scenarios of Easter Island during the last two millennia, indicating the main cultural features discussed in Chapter 2 and summarized in Fig. 2.21. Drought phases are highlighted by *gray* bands. *Blue* lines represent the cultural features depicted in Fig. 2.21 and the bars below indicate the vegetation shifts of each paleoecological site (Aroi, Raraku, Kao). *Black* bars represent charcoal (ch) occurrence and the *brown* bar in Kao represents the presence of coprophilous fungi (cf), especially *Sporormiella*. *Red* arrows indicate possible geographical displacements of the Rapanui population. *Green* arrow marks the European contact (EC) and *blue* arrows indicate the Samalas (S) and Kuwae (K) volcanic eruptions (Fig. 7.1). DACP, Dark Ages Cold Period; LIA, Little Ice Age; MCA, Medieval Climate Anomaly; RWP, Roman Warm Period; EI, Easter Island.

The third step is to select the elements and processes of the subsystems E, H, and L to be analyzed, according to the specific research question under scrutiny. In this case, the environmental factors considered are climate change and geology; the main landscape processes are deforestation, lake desiccation, and soil erosion; and the relevant human developments are demography, the development of new cultivation techniques, and the cultural shift from the moai cult to the Birdman cult (Fig. 8.5). Then, all potential hypotheses derived from this framework should be explicitly formulated by incorporating internal and external feedbacks and synergies between and among these elements. In this case, the account will be subdivided into two

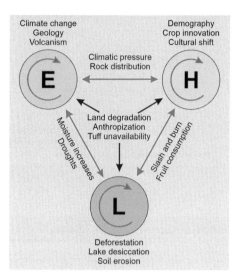

Fig. 8.5 EHLFS holistic approach for Easter Island's prehistoric ecological and cultural changes. *(Redrawn from Rull, V. 2018. Strong Fuzzy EHLFS: a general conceptual framework to address past records of environmental, ecological and cultural change. Quaternary 1, 10.)*

main time intervals, namely the initial development of the Rapanui society (Phase 1) and the shift from the moai cult to the Birdman cult (Phase 2). For this exercise, Figs. 8.4 and 8.5 will be used simultaneously.

Phase 1. Polynesian settlers arrived sometime between 800 and 1200 CE, during the second half of the MCA drought, possibly aided by favorable wind patterns and higher sea levels. However, archaeological evidence of the moai cult is not consistent until about 1200 CE, which is the most accepted date for the archaeological colonization of the island. This is not incompatible with an earlier settlement of small and/or intermittent hunter-gatherer populations, whose eventual archaeological record is negligible or has been cleared by erosion and/or intentional destruction, or hidden by recent rise in the sea level. The moai cult (1200–1600 CE) progressed under wet climates and rising lake levels, and the ancient Rapanui culture established and attained unknown population numbers. The development of the Rapanui society and the moai cult coincided with the anthropogenic island-wide deforestation (with or without the aid of human-introduced rats that actively ate palm fruits), which proceeded at different times and rates across the island. Deforestation typically proceeded by pulses, sometimes followed by regeneration trends, until total forest disappearance, which occurred by 1450 CE in Rano Raraku, 1600 CE in Rano Kao, and 1620 CE in Rano Aroi. Amplification feedbacks between landscape opening and local microclimates would have increased vegetation flammability thus exacerbating fires, and increased soil erosion would have hindered forest recovery. As deforestation progressed, formerly forested areas were progressively replaced by crops within meter-scale stone structures (manavai) that protected soils from erosion, evaporation, and nutrient depletion. Lakeshores were also cultivated using terracing techniques and probably local irrigation methods. During this phase, favorable climates (E), deforestation, and crop innovation (H) interacted among them in a synergistic manner leading to land degradation and significant landscape (L) anthropization. Land degradation by deforestation and soil erosion could have or have not led to resource depletion, internal conflicts, and reductions in human populations. However, the progressive introduction of manavai and lithic mulching techniques, along with spatial reorganizations to take advantage of the more productive soils, may be viewed as an adaptive land-use strategy (internal H feedback) in an island where cultivation was totally dependent on rainfall. As a result, the Rapanui population may have remained relatively stable thus avoiding collapse.

Phase 2. This phase began in the mid-16th century and was likely the consequence of positive feedbacks and synergies between anthropogenic deforestation (H, L) and a climatic drought (E). The deforestation of Rano Raraku (1450 CE) and the desiccation of its internal lake, linked to a multicentennial LIA drought (1570–1720 CE), may have transformed this catchment into dry badlands (L) hard for human life and the maintenance of the moai-carving industry. This could have had several consequences for the Rapanui society, which can also be considered adaptive. First, the search for freshwater for cultivation and human consumption may have favored the displacement of the cultural Rapanui core from Rano Raraku to Rano Kao, with abundant and accessible freshwater. There is no evidence of structures such as channels or aqueducts to transport water across the island, which suggests that the Rapanui lived close to the available freshwater sources. Second, the unavailability of tuff in the Kao crater—and in the whole island—could have contributed to the end of the moai industry. Third, the deep social, religious, and political revolution represented by the Birdman cult has been considered to be more appropriate for unpredictable environmental conditions as those experienced by the Rapanui society after the LIA drought, which must have been very surprising if compared with the former wet climates and environmental stability. In summary, all these feedback and synergies between E, H, and L elements could have been influential in the deep cultural shift experienced by the Rapanui society and its geographical implications.

The EHLFS approach and its application to Easter Island should not be considered a conclusion in itself but a factory of hypotheses to be tested using the MWH and SI methods. Here, a number of hypotheses have been proposed that consider environmental changes reconstructed from paleo-ecological research but others are possible and, hopefully, new proposals will emerge from future research. The main advantages of the EHLFS system are that it avoids a priori determinisms and ruling theories and helps analyzing a complex problem from a holistic perspective. It is also useful to identify specific problems and research questions to guide future research. Finally, the EHLFS approach is well suited to encourage collaboration among different research disciplines, an issue that has been poorly addressed on Easter Island, as explained in the epilogue.

Epilogue: A plea for true interdisciplinarity

A constant in Easter Island's research has been the isolation of different scientific disciplines, which has created a bias in the existing hypotheses according to the field of research in which they have emerged. Some authors integrated evidence from varied disciplines, but biases toward specific ruling theories are evident. A few multidisciplinary teams including paleoecologists and archaeologists have published a number of papers but the research topics are not fully interdisciplinary. On the other hand, archaeologists, anthropologists, ethnographers, and historians, rarely, if ever, refer to paleoecological studies, except for some mentions to Flenley's pioneering works as an introduction to the ecocidal paradigm. Curiously, as we have seen above, it was not Flenley and his team but Mulloy (1974), who proposed (and described in detail) the collapse theory for the first time. The divorce between archaeology and paleoecology has been especially outstanding and has generated harsh debates including personal attacks with explicit political connotations (e.g., Flenley and Bahn, 2007a,b; Hunt and Lipo, 2007). A lesson from this type of controversies is that citations must always be verified in the original sources, rather than accepting wrong statements and unwarranted interpretations of some authors on other's work. Also, care must be taken not to transmit these attitudes to new generations to avoid that young researchers adopt these practices in a natural manner, which may curtail their own free will and, hence, their scientific potential.

The author is unaware of joint initiatives by interdisciplinary teams incorporating paleoecological, archaeological, anthropological, ethnographic, and historical discussions. In this way, each discipline has progressed separately, and the potential benefits of their interaction have not been fully exploited. Given the complexity of the research topics addressed and the interdisciplinary nature of the evidence required for sound explanations, a truly multidisciplinary approach is imperative on Easter Island (Rull et al., 2013). As part of a general call for mutual understanding among scholars from humanities and natural sciences (Förster et al., 2013), the need to establish collaboration between researchers from disparate disciplines working on Easter Island is emphasized, to pave the way toward more general hypotheses that may lead to more holistic views of island's prehistory (Rull, 2018). Paleoecology, archaeology, anthropology, ethnography, history, and related disciplines should be viewed as complementary, rather than excluding sciences whose practitioners compete for an imaginary right, unique, and definitive evidence from their own fields. Each of these disciplines is part of the same puzzle and we should learn how to assemble it together instead of claiming the absolute truth using only our particular, always incomplete, set of pieces. Personal and political issues should not be part of scientific discussions and cites of other's work should be always fair and objective, following the rules of scientific research. The EHLFS approach could be a good start and is able to evolve with future improvements. Integrated interdisciplinary research is needed not only to unravel local historical trends but also to address global concerns. Human-environment interactions have been constant during the recent history and may determine the future trends of our planet. Therefore, transdisciplinarity is a necessity when addressing the potential consequences of future global change using past records as analogs (Van der Leeuw et al., 2011; Dearing, 2013).

References

Abram, N.J., Wright, N.M., Ellis, B., Dixon, B.C., Wurtzel, J.B., England, M.H., Ummenhofer, C.C., Philibosian, B., Cahyarini, S.Y., Yu, T.-L., Shen, C.-C., Cheng, H., Edwards, R.L., Heslop, D., 2020. Coupling of Indo-Pacific climate variability over the last millennium. Nature 579, 385–392.

Alcayaga, S., Narbona, M., 1969. Reconocimiento Detallado de Suelos de la Isla de Pascua. Corporación de Fomento de la Producción, Santiago de Chile.

Allen, M.S., 2006. New ideas about Late Holocene climate variability in the central Pacific. Curr. Anthropol. 47, 521–535.

Amorós, F., 1992. The slave trade run by Spaniards in the Pacific Ocean during the 19th century: the case of Easter Island. Rapa Nui J. 6, 23–26.

Anderson, A., 2002. Faunal collapse, landscape change and settlement history in remote Oceania. World Archaeol. 33, 375–390.

Arnold, M., Orliac, M., Valladas, H.V., 1990. Données nouvelles sur la disparition du Palmier (cf. *Jubaea*) de l'Ile de Pâques. In: Esen-Baur, H.M. (Ed.), State and Perspectives of Scientific Research in Easter Island Culture. 125. Courier Forschungsinstitut Senckenburg, Frankfurt, pp. 217–219.

Ayres, W., 1975. Easter Island: Investigation in Prehistoric Cultural Dynamics. Report submitted to the National Science Foundation, University of South Carolina.

Ayres, W.S., Stevenson, C.P., 2000. Preface. In: Stevenson, C.P., Ayres, W.S. (Eds.), Easter Island Archaeology: Research on Early Rapanui Culture. Easter Island Foundation, Los Osos, pp. ix–xi.

Ayres, W.S., Fitzpatrick, S.M., Wozniak, J., Goleš, G.G., 1998. Archaeological investigations of stone adzes and quarries on Easter Island. In: Stevenson, C.M., Lee, G., Morin, F.J. (Eds.), Easter Island in Pacific Context. Proceedings of the Fourth International Conference on Easter Island and East Polynesia. Easter Island Foundation, Los Osos, pp. 304–311.

Ayres, W.S., Spear, R.L., Beardsley, F.R., 2000. Easter Island obsidian artifacts: typology and use-wear. In: Stevenson, C.P., Ayres, W.S. (Eds.), Easter Island Archaeology: Research on Early Rapanui Culture. Easter Island Foundation, Los Osos, pp. 173–190.

Azizi, G., Flenley, J.R., 2008. The last glacial maximum climatic conditions on Easter Island. Quat. Int. 184, 166–176.

Bahn, P., 1996. Making sense of rongorongo. Nature 379, 205.

Bahn, P., 2015. The end of the moai—did they fall or were they pushed? In: Cauwe, N., De Dapper, M. (Eds.), International Conference Easter Island: Collapse or transformation? A State of the Art. Royal Academy of Overseas Science, Royal Museums of Art and History, Belgian Science Policy Office, Brussels, pp. 135–152.

Bahn, P., Flenley, J., 1992. Easter Island, Earth Island. Tames & Hudson, London.

Baker, P.E., 1998. Petrological factors influencing the utilization of stone on Easter Island. In: Stevenson, C.M., Lee, G., Morin, F.J. (Eds.), Easter Island in Pacific Context. Proceedings of the Fourth International Conference on Easter Island and East Polynesia. Easter Island Foundation, Los Osos, pp. 279–283.

Basener, W., Ross, D., 2005. Booming and crashing populations and Easter Island. SIAM J. Appl. Math. 65, 684–701.

Basener, W., Brooks, B., Radin, M., Wiandt, T., 2008. Rat instigated human population collapse on Easter Island. Nonlin. Dyn. Psychol. Life Sci. 12, 227–240.

Beardsley, R.R., Goleš, G.G., 1998. Sampling the field: provenance studies on prospective rock sources. In: Stevenson, C.M., Lee, G., Morin, F.J. (Eds.), Easter Island in Pacific Context. Proceedings of the Fourth International Conference on Easter Island and East Polynesia. Easter Island Foundation, Los Osos, pp. 284–291.

Birks, H.J.B., 2008. Paleoecology. Encyclopedia of Ecology, vol. 1. Elsevier, Amsterdam, pp. 2623–2634.

Birks, H.J.B., Birks, H.H., 1980. Quaternary Palaeoecology. E. Arnold, London.

Birks, H.J.B., Heiri, O., Seppä, H., Bjune, A.E., 2010. Strenghts and weaknesses of quantitative climate reconstructions based on Late-Quaternary biological proxies. Open Ecol. J. 3, 68–110.

Birks, H.J.B., Lotter, A.F., Juggins, S., Smol, J.P. (Eds.), 2012. Tracking Environmental Change Using Lake Sediments. Data Handling and Numerical Techniques, vol. 5. Kluwer, Dordrecht.

Björk, S., Wohlfarth, B., 2001. ^{14}C chronsostratigraphyc techniques in paleoelimnology. In: Last, W.M., Smol, J.P. (Eds.), Tracking Environmental Change Using Lake Sediments. Basin Analysis, Coring, and Chronological Techniques, vol. 1. Kluwer, Dordrecht, pp. 205–245.

Blaauw, M., 2010. Methods and code for classical age-modelling of radiocarbon sequences. Quat. Geochronol. 5, 512–518.

Blaauw, M., Christen, J.A., 2011. Flexible paleoclimate age-depth models using an autoregressive gamma process. Bayesian Anal. 6, 457–474.

Blaauw, M., Christen, J.A., Bennett, K.D., Reimer, P.J., 2018. Double the dates and go for Bayes—impacts of model choice, dating density and quality on chronologies. Quat. Sci. Rev. 188, 58–66.

Boersema, J.J., 2015a. The Survival of Easter Island. Dwindling Resources and Cultural Resilience. Cambridge University Press, Cambridge.

Boersema, J.J., 2015b. Revisiting the collapse of Rapa Nui (Easter Island) through a voyage of 18th-century journals. In: Cauwe, N., De Dapper, M. (Eds.), International Conference Easter Island: Collapse or Transformation? A State of the Art. Royal Academy of Overseas Science, Royal Museums of Art and History, Belgian Science Policy Office, Brussels, pp. 153–175.

Boersema, J.J., 2017. An earthly paradise? Easter Island (Rapa Nui) as seen by the eighteenth-century European explorers. In: Haoa Cardinali, S., Ingersoll, K.B., Stevenson, C.M. (Eds.), Cultural and Environmental Change on Rapa Nui. Routledge, London, pp. 156–178.

Bork, H.R., Mieth, A., 2003. The key role of *Jubaea* palm trees in the history of Rapa Nui, a provocative interpretation. Rapa Nui J. 17, 119–121.

Bork, H.R., Mieth, A., Tschochner, B., 2004. Nothing but stones? A review of the extent and technical efforts of prehistoric stone mulching on Rapa Nui. Rapa Nui J. 18, 10–14.

Bourke, R.M., 2009. Sweetpotato in Oceania. In: Loebenstein, G., Thottappilly, G. (Eds.), The Sweetpotato. Springer, Berlin, pp. 489–502.

Bovell-Benjamin, A.C., 2007. Sweet potato: a review of its past, present, and future role in human nutrition. Adv. Food Nutr. Res. 52, 1–59.

Bradley, R.S., 2015. Paleoclimatology. Reconstructing Climates of the Quaternary. Academic Press, Oxford.

Bradley, R.S., Briffa, K.R., Cole, J., Hughes, M.K., Osborn, T.J., 2003. The climate of the last millennium. In: Alverson, K.D., Bradley, R.S., Pedersen, T.F. (Eds.), Paleoclimate, Global Change and the Future. Springer, Berlin, pp. 105–141.

Brander, J., Taylor, M., 1998. The simple economics of Easter Island: a Ricardo-Malthus model of renewable resource use. Am. Econ. Rev. 88, 119–138.

Brandt, G., Merico, A., 2015. The slow demise of Easter Island: insights from a model investigation. Front. Ecol. Evol. 3, 13.

Brosnan, T., Becker, M.W., Lipo, C.P., 2018. Coastal groundwater discharge and the ancient inhabitants of Rapa Nui (Easter Island), Chile. Hydrogeol. J. 27, 519–534.

Bull, I.D., Lockheart, M.J., Elhmmali, M.M., Roberts, D.J., Evershed, R.P., 2002. The origin of faeces by means of biomarker detection. Environ. Int. 27, 647–654.

Butler, K.R., Flenley, J.R., 2001. Further pollen evience from Easter Island. In: Stevenson, C.M., Lee, G., Morin, F.J. (Eds.), Pacific 2000. Proceedings of the Fifth International Conference on Easter Island and the Pacific. Easter Island Foundation, Los Osos, pp. 79–86.

Butler, K.R., Flenley, J.R., 2010. The Rano Kau 2 pollen diagram: palaeoecology revealed. Rapa Nui J. 24, 5–10.

Butler, K., Prior, C.A., Flenley, J.R., 2004. Anomalous radiocarbon dates from Easter Island. Radiocarbon 46, 395–420.

Cañellas-Bolta, N., 2014. Vegetation Dynamics in Relation to Climate Changes, Geological Processes and Human Impact at Easter Island Since the Last Glacial. Paleoecological Study of Sediments from Lake Raraku. (Ph.D. dissertation). University of Barcelona, Spain.

Cañellas-Boltà, N., Rull, V., Sáez, A., Margalef, O., Giralt, S., Pueyo, J.J., Birks, H.H., Birks, H.J.B., Pla-Rabes, S., 2012. Macrofossils in Raraku Lake (Easter Island) integrated with sedimentary and geochemical records: towards a palaeoecological synthesis of the last 34,000 years. Quat. Sci. Rev. 34, 113–126.

Cañellas-Boltà, N., Rull, V., Sáez, A., Margalef, O., Bao, R., Pla-Rabes, S., Blaauw, M., Valero-Garcés, B., Giralt, S., 2013. Vegetation changes and human settlement of Easter Island during the last millennia: a multiproxy study of the Lake Raraku sediments. Quat. Sci. Rev. 72, 36–48.

Cañellas-Boltà, N., Rull, V., Sáez, A., Prebble, M., Margalef, O., 2014. First records and potential paleoecological significance of *Dianella* (Xanthorrhoeaceae), an extinct representative of the native flora of Easter Island. Veget. Hist. Archaeobot. 23, 331–338.

Cañellas-Boltà, N., Rull, V., Sáez, A., Margalef, O., Bao, R., Pla-Rabes, S., Valero-Garcés, B., Giralt, S., 2016. Vegetation dynamics at Raraku Lake catchment (Easter Island) during the past 34,000 years. Palaeogeogr. Palaeoclimatol. Palaeoecol. 446, 55–69.

Carlquist, S., 1967. The biota of long-distance dispersal. V. Plant dispersal to Pacific Islands. Bull. Torrey Bot. Club 44, 129–162.

Carlquist, S., 1974. Island Biology. Columbia University Press, New York.

Castañeda, I.S., Schouten, S., 2011. A review of molecular organic proxies for examining modern and ancient lacustrine environments. Quat. Sci. Rev. 30, 2851–2891.

Cauwe, N., Huyge, D., Coupe, D., de Poorter, A., De Dapper, M., de Meulemeester, J., al-Shqour, R., 2007. Ahu Motu Toremo Hiva (Poike Peninsula, Easter Island): dynamic architecture of a series of ahu. In: Wallin, P., Martinsson-Wallin, H. (Eds.), The Gotland Papers. Selected Papers from the VII International Conference on Easter Island and the Pacific. Migration, Identity, and Cultural Heritage. Gotland University Press, Sweden, pp. 47–56.

Caviedes, C.N., Waylen, P.R., 2011. Rapa Nui: a climatically constrained island? Rapa Nui J. 25, 7–23.

Chamberlin, T.C., 1965. The method of multiple working hyptheses. Science 148, 754–759.

Charola, A.E., 1997. Death of a Moai. Easter Island foundation Occassional Paper 4, 50 p.

Charola, A.E., Padgett, A., Lee, G., 1998. Approaches to evaluating deterioration rates of petroglyphs. In: Stevenson, C.M., Lee, G., Morin, F.J. (Eds.), Easter Island in Pacific Context. Proceedings of the Fourth International Conference on Easter Island and East Polynesia. Easter Island Foundation, Los Osos, pp. 326–330.

Church, F., 1998. Uplnd, lowland, citizen, chief: patterns of use-wear from five Easter Island sites. In: Stevenson, C.M., Lee, G., Morin, F.J. (Eds.), Easter Island in Pacific Context. Proceedings of the Fourth International Conference on Easter Island and East Polynesia. Easter Island Foundation, Los Osos, pp. 312–315.

Church, F., Ellis, J.G., 1996. A use-wear analysis of obsidian tools from *Ana Kionga*. Rapa Nui J. 10, 81–88.

CIREN, 2013a. Mapa de Vegetación de Isla de Pascua. Centro de Información de Recursos Naturales. Ministerio de Agricultura, Santiago de Chile. Available at https://www.ciren.cl/. (Accessed November 6, 2019).

CIREN, 2013b. Delimitación Parque Nacional. Centro de Información de Recursos Naturales. Ministerio de Agricultura, Santiago de Chile. Available at https://www.ciren.cl/. (Accessed November 6, 2019).

Ciszewski, A., Ryn, Z., Szelerewicz, M., 2009. The Caves of Easter Island: Underground World of Rapa Nui. Pracownia Kreatywna Bezliku, Kraków.

Clarke, A.C., Burtenshaw, M.K., McLenachan, P.A., Erickson, D.L., Penny, D., 2006. Reconstructing the origins and dispersal of the Polynesian bottle gourd (*Lagenaria siceraria*). Mol. Biol. Evol. 23, 893–900.

Cobb, K.M., Charles, C.D., Cheng, H., Edwards, R.L., 2003. El Niño/Southern Oscillation and tropical Pacific climate during the last millennium. Nature 424, 271–276.

Commendador, A.S., Dudgeon, J.V., Finney, B.P., Fuller, B.T., Esh, K.S., 2013. A stable isotope (δ^{13}C and δ^{15}N) perspective on human diet on Rapa Nui. Am. J. Phys. Anthropol. 152, 173–185.

Commendador, A.S., Finney, B.P., Fuller, B.T., Tromp, M., Dudgeon, J.V., 2019. Multiproxy isotopic analyses of human skeletal material from Rapa Nui: evaluating the evidence from carbonates bulk collagen, and amino acids. Am. J. Phys. Anthropol. 169, 714–729.

CONAF, 1997. Plan de Manejo, Parque Nacional Rapa Nui. Ministerio de Agricultura, Corporación Nacional Forestal, Santiago de Chile.

Cortez, C., Zárate, E., González-Ferrán, O., 2009. Isla de Pascua: impactos y evidencias del tsunami del 22 de mayo de 1960. XII Congreso Geológico Chileno, Santiago de Chile, S2_007.

Cristino, C., Vargas, P., 1999. Ahu Tongariki, Easter Island: chronological and sociopolitical significance. Rapa Nui J. 13, 67–69.

Cristino, C., Vargas, P., 2002. Archaeological excavations and reconstruction of Ahu Tongariki—Easter Island. Rev. Urban. 5, 1–10.

Cristino, C., Vargas, P., Izaurieta, R., 1981. Atlas Arqueológico de Isla de Pascua. Centro de Estudios Isla de Pascua. Universidad de Chile, Santiago de Chile.

Cummings, L.S., 1998. A review of recent pollen and phytolith studies from various contexts on Easter Island. In: Stevenson, C.M., Lee, G., Morin, F.J. (Eds.), Easter Island in Pacific Context. Proceedings of the Fourth International Conference on Easter Island and East Polynesia. Easter Island Foundation, Los Osos, pp. 100–106.

D'Anjou, R.M., Bradley, R.S., Balascio, N.L., Finkelstein, D.B., 2012. Climate impacts on human settlement and agricultural activities in northern Norway revealed hrough sediment biogeochemistry. Proc. Natl. Acad. Sci. U. S. A. 109, 20332–20337.

Davis, M.B., 1981. Quaternary history and the stability of forest communities. In: West, D.C., Shugart, D.B., Botkin, D.B. (Eds.), Forest Succession, Concepts and Applications. Springer, New York, pp. 132–153.

De Boer, E.J., Rull, V., van Leeuwen, J.F.N., Amaral-Zettler, L.A., Bao, R., Benavente-Marín, M., Gonçalves, V., Hernández, A., Marques, H., Pimentel, C., Pla-Rabes, S., Raposeiro, P.M., Richter, N., Ritter, C., Rubio-Inglés, M.J., Sáez, A., Trigo, R.M., Vázquez-Loureiro, D., Vilaverde, J., Giralt, S., 2019. Ecosystem impacts of human arrival in the Azores: a comparative study of high-resolution multi-proxy lake sediment records. 20th Congress of the International Union of Quaternary Research (INQUA), Dublin, Ireland, O-3174.

Dearing, J.A., 2013. Why future earth needs lake sediment studies. J. Paleolimnol. 49, 537–545.

Delhon, C., Orliac, C., 2007. The vanished palm trees of Easter Island: new radiocarbon and phytolith data. In: Wallin, P., Martinsson-Wallin, H. (Eds.), The Gotland Papers. Selected Papers from the VII International Conference on Easter Island and the Pacific. Migration, Identity, and Cultural Heritage. Gotland University Press, Sweden, pp. 97–110.

DeMenocal, P.B., 2001. Cultural responses to climate change during the Late Holocene. Science 292, 667–673.

DeMets, C., Gordon, R.G.,Vogt, P., 1994. Location of the Africa-Australia-India tripple junctionand motion between the Australian and Indian plates—Results from an aeromagnetic investigation of the Central Indian and Carlsberg Ridges. Geophys. J. Int. 119, 893–930.

Diamond, J., 2005. Collapse. How societies choose to fail or survive. Allen Lane, London.

DiNapoli, R.J., Lipo, C.P., Brosnan, T., Hunt, T.L., Hixon, S., Morrison, A.E., Becker, M., 2018. Rapa Nui (Easter Island) monument (ahu) locations explained by freshwater sources. PLoS One 14, e0210409.

DiNapoli, R.J., Rieth, T.M., Lipo, C.P., Hunt, T.L., 2020. A model-based approach to the tempo of "collapse": the case of Rapa Nui (Easter Island). J. Archaeol. Sci. 116, 105094.

Doswell, C.A., 2002. In the line of fire: first global lightning map reveals high-strike zones. Nat. Geogr. Mag. 202, vii.

Dransfield, J., Flenley, J.R., King, S.M., Harkness, D.D., Rapu, S., 1984. A recently extinct palm from Easter Island. Nature 312, 750–752.

Dubois, A., Lenne, P., Nahoe, E., Rauch, M., 2013. Plantas de Rapa Nui. Guía Ilustrada de la Flora de Interés Ecológico y Patrimonial. Umanga mo te Natura, CONAF, ONF International, Santiago de Chile.

Dumont, H.J., 2002. Impoverished freshwater fauna of Easter Island. Rapa Nui J. 16, 29–30.

Dumont, H.J., Cocquyt, C., Fortugne, M., Arnold, M., Reyss, J.L., Bloemendal, J., Oldfield, F., Steenbergen, C.L.M., Korthals, H.J., Zeeb, B.A., 1998. The end of moai quarrying and its effect on Lake Rano Raraku, Easter Island. J. Paleolimnol. 20, 409–422.

Edwards, E., Edwards, A., 2013. When the Universe was an Island. Exploring the Cultural and Spiritual Cosmos of Ancient Rapanui. Hangaroa Press, Hanga Roa.

Eglington, T.I., Eglington, G., 2008. Molecular proxies for paleoclimatology. Earth Planet. Sci. Lett. 275, 1–16.

Elix, J.A., McCarthy, P.M., 1998. Catalogue of the lichens of the smaller Pacific islands. Bibl. Lichenol. 70, 1–361.

Englert, S., 1948. La Tierra de Hotu Matu'a: Historia y Etnología de la Isla de Pascua. San Francisco, Santiago de Chile.

Englert, S., 1970. Island at the Center of the World. Scribners, New York.

Espejo, J., Rodriguez, F., 2013. New insights in conservation of *Sophora toromiro* (Phil.) Skottsb., emblematic species of the South Pacific. Rapa Nui J. 27, 5–9.

Etienne, M., Michea, G., Díaz, E., 1982. Flora, vegetación y potencial pastoral de la Isla de Pascua. Facultad de Ciencias Agrarias,Veterinarias y Forestales Universidad de Chile, Santiago de Chile.

Faegri, K., Kaland, P.E., Krywinski, K., 1989. Textbook of Pollen Analysis. Wiley, Chichester.

Fagan, B., 2008. The Great Warming: Climate Change and the Rise and Fall of Civilizations. Bloomsbury, New York.

Fehren-Schmitz, L., Jarman, C.L., Harkins, K.M., Kayser, M., Popp, B.N., Skogund, P., 2017. Genetic ancestry of Rapanui before and after European contact. Curr. Biol. 27, 3209–3215.

Ferdon, E.N., 1961. The ceremonial site of 'Orongo. In: Heyerdahl, T., Ferdon, E. (Eds.), Reports of the Norwegian Archaeological Expedition to Easter Island and the East Pacific. Archaeology of Easter Island, vol. 1. Allen & Unwin, London, pp. 221–254.

Figueroa, G., 1979. William Mulloy: 1917-1978. Asian Perspect. 22, 101–105.

Finot, V.L., Marticorena, C., Marticorena, A., Rojas, G., Barrera, J.A., 2015. Grasses (Poaceae) of Easter Island—native and introduced species. In: Blanco, J. (Ed.), Biodiversity in Ecosystems—Linking Structure and Function. IntechOpen, London, https://doi.org/10.5772/59154.

Fischer, S.R., 1997. Rongorongo. The Easter Island Script. Oxford University Press, Oxford.

Fischer, S.R., 2005. Island at the End of the World. The Turbulent History of Easter Island. Reaktion Books, London.

Flenley, J.R., 1979a. The Equatorial Rain Forest: A Geological History. Butterworths, London.

Flenley, J.R., 1979b. Stratigraphic evidence of environmental change on Easter Island. Asian Perspect. 22, 33–40.

Flenley, J.R., 1996. Further evidence of vegetational change on Easter Island. South Pacific St. 16, 135–141.

Flenley, J.R., 1998. New data and new thoughts about Rapa Nui. In: Stevenson, C.M., Lee, G., Morin, F.J. (Eds.), Easter Island in Pacific Context. Proceedings of the Fourth International Conference on Easter Island and East Polynesia. Easter Island Foundation, Los Osos, pp. 125–128.

Flenley, J.R., 2007. A palynologist looks at the colonization of the Pacific. In: Wallin, P., Martrinsson-Wallin, H. (Eds.), The Gotland Papers. Selected Papers from the VII International Conference on Easter Island and the Pacific. Migration, Identity, and Cultural Heritage. Gotland University Press, Sweden, pp. 15–34.

Flenley, J.R., Bahn, P.G., 2003. The Enigmas of Easter Island. Oxford University Press, Oxford.

Flenley, J., Bahn, P., 2007a. Conflicting views of Easter Island. Rapa Nui J. 21, 11–13.

Flenley, J., Bahn, P., 2007b. Respect versus contempt for evidence: reply to Hunt and Lipo. Rapa Nui J. 21, 98–104.

Flenley, J.R., Bahn, P.G., 2011. Hunt, Terry and Carl Lipo. The Statues that Walked. Unravelling the Mystery of Easter Island. Rapa Nui J. 25, 60–62.

Flenley, J.R., King, S., 1984. Late Quaternary pollen records from Easter Island. Nature 307, 47–50.

Flenley, J.R., King, A.S.M., Jackson, J., Chew, C., Teller, J.T., Prentice, M.E., 1991. The Late Quaternary vegetational and climatic history of Easter Island. J. Quat. Sci. 6, 85–115.

Flint, R.F., 1971. Glacial and Quaternary Geology. Wiley, New York.

Flores, J.P., Torres, C., Martínez, E., Muñoz, P., 2013. Determinación de la Erosión Actual y Potencial de los Suelos en la Isla de Pascua. Ministerio de Agricultura, Centro de Información de Recursos Naturales, Santiago de Chile.

Follmann, G., 1961. Lichenometrische Alterbestimmungen an vorchristligen Steinsetzungen der polynesischen Osterinsel. Naturwissenschaften 19, 627–628.

Förster, F., Großmann, R., Hinz, M., Iwe, K., Kinkel, H., Larsen, A., Lungershausen, U., Matarese, C., Meurer, P., Nelle, O., Robin, V., Teichmann, M., 2013. Towards mutual undestranding within interdisciplinary palaeoenvironmental research: an exemplary analysis of the term *landscape*. Quat. Int. 312, 4–11.

Garreaud, R., Aceituno, P., 2001. Interannual rainfall variability over the South American Altiplano. J. Clim. 14, 2779–2789.

Gates, W.L., 1976. Modelling the Ice-Age climate. Science 191, 1138–1141.

Genz, J., Hunt, T.L., 2003. El Niño/southern oscillation and Rapa Nui prehistory. Rapa Nui J. 17, 7–14.

Gill, G.W., 1998. Easter Island settlement: current evidence and future research directions. In: Stevenson, C.M., Lee, G., Morin, F.J. (Eds.), Easter Island in Pacific Context. Proceedings of the Fourth International Conference on Easter Island and East Polynesia. Easter Island Foundation, Los Osos, pp. 137–142.

Gioncada, A., González-Ferran, O., Lezzerini, M., Mazzuoli, R., Bisson, M., Rapu, S.A., 2010. The volcanic rocks of Easter Island (Chile) and their use for the moai sculptures. Eur. J. Mineral. 22, 855–867.

Glew, J.R., Smol, J.P., Last, W.M., 2001. Sediment core collection and extrussion. In: Last, W.M., Smol, J.P. (Eds.), Tracking Environmental Change Using Lake Sediments. Basin Analysis, Coring, and Chronological Techniques, vol. 1. Kluwer, Dordrecht, pp. 73–105.

Glynn, P.W., Wellington, G.M., Wieters, E.A., Navarrete, S.A., 2003. Reef-building communities of Easter Island (Rapa Nui), Chile. In: Cortés, J. (Ed.), Latin American Coral Reefs. Elsevier, Amsterdam, pp. 473–494.

Golson, J., 1965. Thor Heyerdahl and the prehistory of Easter Island. Oceania 36, 38–83.

González-Ferrán, O., Mazzuoli, R., Lahsen, A., 2004. Geología del complejo volcánico Isla de Pascua, Rapa Nui, Chile, V Región Valparaíso. Carta Geológica-Volcánica Isla de Pascua. Centro de Estudios Volcanológicos, Santiago de Chile.

Good, D.H., Reuveny, R., 2006. The fate of Easter Island: the limits of resource management institutions. Ecol. Econ. 58, 473–490.

Goodfellow, B.J., Wilson, R.H., 1990. A Fourier transform IR study of the gelation of amylose and amylopectin. Biopolymers 30, 1183–1189.

Goodwin, D., Browning, S.A., Anderson, A.J., 2014. Climate windows for Polynesian voyaging to New Zealand and Easter Island. Proc. Natl. Acad. Sci. U. S. A. 111, 14716–14721.

Gossen, C., 2007. Report: the mystery lies in the *Scirpus*. Rapa Nui J. 21, 105–110.

Gossen, C., 2011. Deforestation, Drought and Humans: New Discoveries of the Late Quaternary Paleoenvironment of Rapa Nui (Easter Island). (Ph.D. dissertation). Portland State University, Portland.

Gossen, C., Stevenson, C.M., 2005. Prehistoric solar innovation and water management on Rapa Nui. In: International Solar Energy Society/American Solar Energy Society Conference Proceedings. Solar World Congress 4, pp. 2580–2585.

Goudie, A., 2006. The Human Impact on the Natural Environment. Blackwell, Oxford.

Grau, J., 1998. The *Jubaea* palm, key in the transportation of Moai on Easter Island. In: Stevenson, C.M., Lee, G., Morin, F.J. (Eds.), Easter Island in Pacific Context. Proceedings of the Fourth International Conference on Easter Island and East Polynesia. Easter Island Found, Los Osos, pp. 120–124.

Grau, J., 2001. More about *Jubaea chilensis* on Easter Island. In: Stevenson, C.M., Lee, G., Morin, F.J. (Eds.), Pacific 2000. Proceedings of the Fifth International Conference on Easter Island and the Pacific. Easter Island Foundation, Los Osos, pp. 87–90.

Green, R.C., 2000a. Origin for the Rapanui of Easter Island before European contact: solutions from holistic anthropology to an issue no longer much of a mystery. Rapa Nui J. 14, 71–76.

Green, R.C., 2000b. A range of disciplines support a dual origin for the bottle gourd in the Pacific. J. Polynesian Soc. 109, 191–197.

Green, R.C., 2005. Sweet potato transfer in Polynesian prehistory. In: Ballard, C., Brown, P., Nourke, R.M., Harwood, T. (Eds.), The Sweet Potato in Oceania: A Reappraisal. University of Sydney, Sydney, pp. 43–62.

Grolle, R., 2002. The Hepaticae of Easter Island (Chile). Bryologist 105, 126–127.

Grove, R., Adamson, G., 2018. El Niño chronology and the Little Ice Age. In: Grove, R., Adamson, G. (Eds.), El Niño in World History. Palmgrave Macmillan, London, pp. 49–79.

Haase, K.M., Devey, C.W., 1996. Geochemistry of lavas from the Ahu and Tupa volcanic fields, Easter Hotspot, southeast Pacific: implications for intraplate magma genesis near a spreading axis. Earth Planet. Sci. Lett. 137, 129–143.

Haberle, S.G., Chepstow-Lusty, A., 2000. Can climate influence cultural development? A view through time. Environ. Hist. 6, 349–369.

Harrison, A.A., 2011. Fear, pandemonium, equanimity and delight: human responses to extra-terrestrial life. Philos. Trans. R. Soc. A 369, 656–668.

Hather, J., Kirch, P.V., 1991. Prehistoric sweet potato (*Ipomoea batatas*) from Mangaia Island, central Polynesia. Antiquity 65, 887–893.

Haug, G., Günther, D., Peterson, L.C., Sigman, D.M., Hughen, K.A., Aeschlimann, B., 2003. Climate and the collapse of Maya civilization. Science 299, 1731–1735.

Herrera, C., Custodio, E., 2008. Conceptual hydrogeological model of volcanic Easter Island (Chile) after chemical and isotopic surveys. Hydrogeol. J. 16, 1329–1348.

Heyerdahl, T., 1952. American Indians in the Pacific: The Theory Behind the Kon-Tiki Expedition. Allen & Unwin, London.

Heyerdahl, T., 1989. Easter Island. The Mystery Solved. Random House, New York.

Heyerdahl, T., Ferdon, E., 1961. Reports of the Norwegian Archaeological Expedition to Easter Island and the East Pacific. Archaeology of Easter Island, vol. 1. Rand McNally, New York.

Heyerdahl, T., Skjølvold, A., Pavel, P., 1989. The 'walking' *moai* of Easter Island. Occ. Pap. Kon-Tiki Mus. 1, 55.

Higgins, H.G., Stewart, C.M., Harrington, K.J., 1961. Infrared spectra of cellulose and related polysaccharides. J. Polym. Sci. 51, 59–84.

Hill, A., 1981. Why study paleoecology? Nature 293, 340.

Hixon, S.W., Lipo, C.L., McMorran, B., Hunt, T.L., 2018. The colossal hats (*pukao*) of monumental statues on Rapa Nui (Easter Island, Chile): analyses of *pukao* variability, transport, and emplacememnt. J. Archaeol. Sci. 100, 148–157.

Hofreiter, M., Collins, M., Stewart, J.R., 2012. Ancient biomolecules in Quaternary palaeoecology. Quat. Sci. Rev. 33, 1–13.

Hogg, A.G., Hua, Q., Blackwell, P.G., Niu, M., Buck, C.E., Childerson, T.P., Heaton, T.J., Palmer, J.G., Reimer, P.J., Reimer, R.W., Turney, C.S.M., Zimmerman, R.H., 2013. SHCal13 southern hemisphere calibration, 0-50,000 years cal BP. Radiocarbon 55, 1889–1903.

Horley, P., 2007. Comparative structural analysis of *rongorongo* script and Rapa Nui songs. In: Wallin, P., Martrinsson-Wallin, H. (Eds.), The Gotland Papers. Selected Papers from the VII International Conference on Easter Island and the Pacific. Migration, Identity, and Cultural Heritage. Gotland University Press, Sweden, pp. 217–224.

Horrocks, M., Wozniak, J.A., 2008. Plant microfossil analysis reveals disturbed forest and a mixed-crop, dryland production system at Te Niu, Easter Island. J. Archaeol. Sci. 35, 126–142.

Horrocks, M., Baisden, W.T., Flenley, J., Feek, D., González-Nualart, L., Haoa-Cardinali, S., Edmunds Gorman, T., 2012a. Fossil plant remains at Rano Raraku, Easter Island's statue quarry: evidence for past elevated lake level and ancient Polynesian agriculture. J. Paleolimnol. 46, 767–783.

Horrocks, M., Baisden, W.T., Nieuwoudt, M.K., Flenley, J., Feek, D., González Nualart, L., Haoa-Cardinali, S., Edmunds Gorman, T., 2012b. Microfossils of Polynesian cultigens in lake sediment cores from Rao Kau, Easter Island. J. Paleolimnol. 47, 185–204.

Horrocks, M., Marra, M., Baisden, W.T., Flenley, J., Feek, D., González-Nualart, L., Haoa-Cardinali, S., Edmunds Gorman, T., 2013. Pollen, phytoliths, arthropods and high-resolution ^{14}C sampling from Rano Kau, Easter Island: evidence for Late Quaternary environments, ant (formicidae) distributions and human activity. J. Paleolimnol. 50, 417–432.

Horrocks, M., Baisden, W.T., Harper, M.A., Marra, M., Flenley, J., Feek, D., Haoa-Cardinali, S., Keller, E.D., González Nualart, L., Edmunds Gorman, T., 2015. A plant microfossil record of Late Quaternary environments and human activity from Rano Aroi and surroundings, Easter Island. J. Paleolimnol. 54, 279–303.

Horrocks, M., Baisden, T., Flenley, J., Feek, D., Love, C., Haoa-Cardinali, S., González Nualart, L., Edmunds Gorman, T., 2016. Pollen, phytolith and starch analyses of dryland soils from Easter Island (Rapa Nui) show widespread vegetation clearance and Polynesian-introduced crops. Palynology 41, 339–350.

Houlton, B.Z., Bai, E., 2009. Imprint of denitrifying bacteria on the global terrestrial biosphere. Proc. Natl. Acad. Sci. U. S. A. 106, 21713–21716.

Hubbard, D.K., García, M., 2003. The corals and coral reefs of Easter Islad—a preliminary look. In: Loret, J., Tanacredi, J.T. (Eds.), Easter Island. Scientific Exploration into the World's Environmental Problems in Microcosm. Kluwer Academic/Plenum, New York, pp. 53–77.

Hunt, T.L., 2006. Rethinking the fall of Easter Island. New evidence points to and alternative explanation for a civilization's collapse. Am. Sci. 94, 412–419.

Hunt, T.L., 2007. Rethinking Easter Island's ecological catastrophe. J. Archaeol. Sci. 34, 485–502.

Hunt, T.L., Lipo, C.P., 2006. Late colonization of Easter Island. Science 311, 1603–1606.

Hunt, T.L., Lipo, C.P., 2007. Chronology, deforestation, and "collapse": evidence vs. faith in Rapa Nui prehistory. Rapa Nui J. 21, 85–97.

Hunt, T.L., Lipo, C.L., 2009. Revisiting Rapa Nui (Easter Island) "ecocide". Pac. Sci. 63, 601–616.

Hunt, T.L., Lipo, C., 2011. The Statues that Walked. Free Press, New York.

Hunt, T.L., Lipo, C.P., 2018. The archaeology of Rapa Nui (Easter Island). In: Cochrane, E.E., Hunt, T.L. (Eds.), The Oxford Handbook of Prehistoric Oceania. Oxford University Press, Oxford, pp. 416–449.

Hunter-Anderson, R.L., 1998. Human vs. climatic impacts at Rapa Nui, did people really cut down all those trees? In: Stevenson, C.M., Lee, G., Morin, F.J. (Eds.), Easter Island in Pacific Context. Proceedings of the Fourth International Conference on Easter Island and East Polynesia. Easter Island Foundation, Los Osos, pp. 95–99.

Ingersoll, K.B., Ingersoll, D.W., 2017. The potential for palm extinction on Rapa Nui by disease. In: Haoa Cardinali, S., Ingersoll, K.B., Stevenson, C.M. (Eds.), Cultural and Environmental Change on Rapa Nui. Routledge, London, pp. 59–73.

Ingersoll, K.B., Ingersoll, D.W., Bove, A., 2017. Healing a culture's reputation. Challenging the cultural labeling and libeling of the Rapanui. In: Haoa Cardinali, S., Ingersoll, K.B., Stevenson, C.M. (Eds.), Cultural and Environmental Change on Rapa Nui. Routledge, London, pp. 188–202.

Ireland, R.R., Bellolio, G., 2002. The mosses of Easter Island. Trop. Briol. 21, 11–19.

Jalut, G., Deboubat, J.J., Fortugne, M., Otto, T., 2009. Holocene circum-Mediterranean vegetation changes: climate forcing and human impact. Quat. Int. 200, 4–18.

Jarman, C.L., Larsen, T., Hunt, T.L., Lipo, C.P., Solsvik, R., Wallsgrove, N., Ka'apu-Lyons, C., Close, H.G., Popp, B.N., 2017. Diet of the prehistoric population of Rapa Nui (Easter Island, Chile) shows environmental adaptation and resilience. Am. J. Phys. Anthropol. 164, 343–361.

Jones, P.D., Osborn, T.J., Briffa, K.R., 2001. The evolution of climate over the last millennium. Science 292, 662–667.

Jones, T.L., Storey, A.A., Matisoo-Smith, E., Ramírez-Aliaga, J.M., 2011. Polynesians in America. Pre-Columbian Contact with the New World. Altamira Press, Landham.

Jowsey, P.C., 1966. An improved peat sampler. New Phytol. 65, 245–248.

Junk, C., Claussen, M., 2011. Simulated climate variability in the region of Rapa Nui during the last millennium. Clim. Past 7, 579–586.

Kirch, P.V., 1984. The Evolution of the Polynesian Chiefdoms. Cambridge University Press, Cambridge.

Kirch, P.V., 2010. Peopling of the Pacific: a holistic anthropological perspective. Annu. Rev. Anthropol. 39, 131–148.

Kirch, P.V., Ellison, J., 1994. Palaeoenvironmental evidence for human colonisation of remote Oceanic islands. Antiquity 68, 310–321.

Kirch, P.V., Green, R., 1987. History, phylogeny and evolution in Polynesia. Curr. Anthropol. 28, 431–456.

Köppen, W., Wegener, A., 1924. Die Klimate der geologischen Vorzeit. Gebrüder Borntraeger, Berlin.

La Pérouse, J.F.G., 1797. Voyage de La Pérouse Autour du Monde. Imprimerie de la République, Paris.

Ladefoged, T., Stevenson, C.M., Vitousek, P., Chadwick, O., 2005. Soil nutrient depletion and the collapse of Rapa Nui society. Rapa Nui J. 19, 100–105.

Ladefoged, T.N., Stevenson, C.M., Haoa, S., Mulrooney, M., Puleston, C., Vitousek, P.M., Chadwick, O.A., 2010. Soil nutrient analysis of Rapa Nui gardening. Archaeol. Ocean. 45, 80–85.

Ladefoged, T.N., Flaws, A., Stevenson, C.M., 2013. The distribution of rock gardens on Rapa Nui (Easter Island) as determined from satellite imagery. J. Archaeol. Sci. 40, 1203–1212.

Last, W.M., Smol, J.P. (Eds.), 2001a. Tracking Environmental Change Using Lake Sediments. Basin Analysis, Coring, and Chronological Techniques, vol. 1. Kluwer, Dordrecht.

Last, W.M., Smol, J.P. (Eds.), 2001b. Tracking Environmental Change Using Lake Sediments. Physical and Geochemical Methods, vol. 2. Kluwer, Dordrecht.

Lavachery, H.A., 1939. Les Pétroglyphes de l'Île de Pâques. De Sikkel, Antwerp.

Lawler, A., 2010. Beyond the Kon-Tiki: did Polynesians sail to South America? Science 328, 1344–1347.

Lee, G., 1992. The Rock Art of Easter Island. Symbols of Power, Prayers to the Goods. Institute of Archaeology, University of California, Los Angeles.

Lee, G., Liller, W., 1987. Easter Island's "sun stones": a re-evaluation. Archaeoastronomy 11, S1–S11.

Lee, G., Bahn, P., Horley, P., Haoa Cardinali, S., Gozález Nualart, L., Cuadros Hucke, N., 2017. Re-use of the sacred. Late period petroglyphs applied to red scoria topknots from Easter Ilsand (Rapa Nui). In: Haoa Cardinali, S., Ingersoll, K.B., Stevenson, C.M. (Eds.), Cultural and Environmental Change on Rapa Nui. Routledge, London, pp. 133–155.

Lewis, C., Costa, F.R.C., Bongers, F., Peña-Claros, M., Clement, C.R., Junqueira, A.B., Neves, E.G., Tamanaha, E.K., Figueiredo, F.O.G., Salomão, R.P., et al., 2017. Persistent effects of pre-Columbian plant domestication on Amazon forest composition. Science 355, 925–931.

Likens, G.E., Davis, M.B., 1975. Post-glacial history of Mirror Lake and its watershed in New Hampshire, U.S.A., an initial report. Int. Ver. Theor. Angew. Limnol. Verh. 19, 982–993.

Lipo, C.P., Hunt, T.L., 2005. Mapping prehistoric statue roads on Easter Island. Antiquity 79, 158–168.

Lipo, C.P., Hunt, T.L., 2016. Chronology and Easter Island prehistory. In: Stefan, V.H., Gill, G.W. (Eds.), Skeletal Biology of the Ancient Rapanui (Easter Islanders). Cambridge University Press, Cambridge, pp. 39–65.

Lipo, C.P., Hunt, T.L., Rapu, S., 2013. The 'walking' megalithic statues (*moai*) of Easter Island. J. Archaeol. Sci. 40, 2859–2866.

Lipo, C.P., Hunt, T.L., Horneman, R., 2016. Weapons of war? Rapa Nui mata'a morphometric analyses. Antiquity 90. 172.187.

Lipo, C.P., DiNapoli, R.J., Hunt, T.L., 2018. Commentary: rain, sun, soil, and sweat: a consideration of population limits on Rapa Nui (Easter Island) before European contact. Front. Ecol. Evol. 6, 25.

Litt, T., Gibbard, P., 2008. Definition of a global stratotype section and point (GSSP) for the base of the upper (Late) Pleistocene Subseries (Quaternary System/Period). Episodes 31, 260–263.

Lotka, A.J., 1925. Elements of Physical Biology. Williams & Wilkins, Baltimore.

Love, C.M., 1990. How to make and move an Easter Island statue. In: Esen-Baur, H.M. (Ed.), State and Perspectives of Scientific Research in Easter Island Culture, vol. 125. Courier Forschungsinstitut Senckenburg, Frankfurt, pp. 139–140.

Love, C.M., 2007. The Easter Island cultural collapse. In: Wallin, P., Martrinsson-Wallin, H. (Eds.), The Gotland Papers. Selected Papers from the VII International Conference on Easter Island and the Pacific. Migration, Identity, and Cultural Heritage. Gotland University Press, Sweden, pp. 67–86.

MacIntyre, F., 2001. ENSO, climate variability and the Rapanui. Part II, oceanography and the Rapanui. Rapa Nui J. 15, 83–94.

Mahon, I., 1998. Easter Island: the economics of population dynamics and sustainable development in Pacific context. In: Stevenson, C.M., Lee, G., Morin, F.J. (Eds.), Easter Island in Pacific Context. Proceedings of the Fourth International Conference on Easter Island and East Polynesia. Easter Island Foundation, Los Osos, pp. 113–119.

Malthus, T.R., 1798. An Essay on the Principle of Population, as it Affects the Future Improvement of Society, with Remarks on the Speculations of Mr. Godwin, Mr. Condorcet, and Other Writers. J. Johnson, London.

Mann, D., Chase, J., Edwards, J., Beck, W., Reanier, R., Mass, M., 2003. Prehistoric destruction of the primeval soils and vegetation of Rapa Nui (Isla de Pascua, Easter Island). In: Loret, J., Tanacredi, J.T. (Eds.), Easter Island. Scientific Exploration into the World's Environmental Problems in Microcosm. Kluwer Academic/Plenum, New York, pp. 133–153.

Mann, D., Edwards, J., Chase, J., Beck, W., Reanier, R., Mass, M., Finney, B., Loret, J., 2008. Drought, vegetation change, and human history on Rapa Nui (Isla de Pascua, Easter Island). Quat. Res. 69, 16–28.

Margalef, O., 2014. The Last 70 kyr of Rano Aroi (Easter Island, 27° S) Peat Record: New Insights for the Central Pacific Paleoclimatology. (Ph.D. dissertation). University of Barcelona, Spain.

Margalef, O., Cañellas-Boltà, N., Pla-Rabes, S., Giralt, S., Pueyo, J.J., Joosten, H., Rull, V., Buchaca, T., Hernández, A., Valero-Garcés, B.L., Moreno, A., Sáez, A., 2013. A 70,000 year geochemical and palaeoecological record of climatic and environmental change from Rano Aroi peatland (Easter Island). Glob. Planet. Chang. 108, 72–84.

Margalef, O., Martínez-Cortizas, A., Kylander, M., Pla-Rabes, S., Cañellas-Boltà, N., Pueyo, J.J., Sàez, A., Valero-Garcés, B., Giralt, S., 2014. Environmental processes in Rano Aroi (Easter Island) peat geochemistry forced by climate variability during the last 70 kyr. Palaeogeogr. Palaeoclimatol. Palaeoecol. 414, 438–450.

Margalef, O., Álvarez-Gómez, J.S., Pla-Rabes, S., Cañellas-Boltà, N., Rull, V., Sáez, A., Geyer, A., Peñuelas, J., Sardans, J., Giralt, S., 2018. Revisiting the role of high-enegry Pacific events in the environmental and cultural history of Easter Island (Rapa Nui). Geogr. J. 184, 310–322.

Martinsson-Wallin, H., 1994. Ahu-The Ceremonial Stone Structures of Easter Island. Societas Archaeologica Upsalensis, Uppsala.

Martinsson-Wallin, H., 1998. Excavations at Ahu Heki'i, La Pérouse, Easter Island. In: Stevenson, C.M., Lee, G., Morin, F.J. (Eds.), Easter Island in Pacific Context. Proceedings of the Fourth International Conference on Easter Island and East Polynesia. Easter Island Foundation, Los Osos, pp. 171–177.

Martinsson-Wallin, H., 2001. Construction—destruction—reconstruction of monumental architerchture on Rapa Nui. In: Stevenson, C.M., Lee, G., Morin, F.J. (Eds.), Pacific 2000. Proceedings of the Fifth International Conference on Easter Island and the Pacific. Easter Island Foundation, Los Osos, pp. 73–77.

Martinsson-Wallin, H., Crockford, S.J., 2002. Early settlement of Rapa Nui (Easter Island). Asian Perspect. 40, 244–278.

Martinsson-Wallin, H., Wallin, P., 2000. Ahu and settlement: archaeological excavations at 'Anakena and La Pérouse. In: Stevenson, C.P., Ayres, W.S. (Eds.), Easter Island Archaeology: Research on Early Rapanui Culture. Easter Island Foundation, Los Osos, pp. 27–44.

Martinsson-Wallin, H., Wallin, P., 2014. Spatial perspectives on ceremonial complexes: testing traditional land divisions on Rapa Nui. Stdud. Global Archaeol. 20, 317–342.

Maunder, M., Culham, A., Alden, B., Zizka, G., Orliac, C., Lobin, W., Bordeu, A., Ramírez, J.M., Glissmann-Gough, S., 2000. Conservation of the toromiro tree, case study in the management of a plant extinct in the wild. Conserv. Biol. 14, 1341–1350.

McCall, G., 1981. Rapanui. Tradition and Survival on Easter Island. University Press Hawaii, Honolulu.

McCall, G., 1993. Little Ice Age, some speculations for Rapanui. Rapa Nui J. 7, 65–70.

McCall, G., 2009. Easter Island. In: Gillespie, R.G., Clague, D.A. (Eds.), Encyclopedia of Islands. University of California Press, Berkeley, pp. 244–251.

McCormac, F.G., Hogg, A.G., Blackwell, P.G., Buck, C.E., Higham, T.F.G., Reimer, P.J., 2004. ShCal04 Southern Hemisphere calibration, 0-11.0 cal kyr BP. Radiocarbon 46, 1087–1092.

McCoy, P.C., 1976. Easter Island Settlement Patterns in the Late Prehistoric and Protohistoric Periods. Bull. 5, International Fund for Monuments. Easter Island Committee, New York.

McCoy, P.C., 1979. Easter Island. In: Jennings, J.D. (Ed.), The Prehistory of Polynesia. Harvard University Press, Cambridge, pp. 135–166.

McGlone, M.S., 1983. Polynesian deforestation of New Zealand: a preliminary synthesis. Archaeol. Ocean. 18, 11–25.

McLauchlan, K.K., Williams, J.J., Craine, J.M., Jeffers, E.S., 2013. Changes in nitrogen cycling during the Holocene epoch. Nature 495, 352–355.

McLaughlin, S., 2007. The Complete Guide to Easter Island. Easter Island Foundation, Los Osos.

McMichael, C.N.H., Mattews-Bird, F., Farfan-Rios, W., Feeley, K.J., 2017. Ancient human disturbances may be skewing our understanding of Amazon forests. Proc. Natl. Acad. Sci. U. S. A. 114, 522–527.

Merico, A., 2017. Models of Easter Island human-resource dynamics: advances and gaps. Front. Ecol. Evol. 5, 154.

Métraux, A., 1940. Ethnology of Easter Island. Bishop Museum, Honolulu.

Meyer, J.Y., 2013. A note on the taxonomy, ecology, distribution and conservation status of the ferns (Pteridophytes) of Rapa Nui (Easter Island). Rapa Nui J. 27, 71–83.

Mieth, A., Bork, H.R., 2003. Diminution and degradation of environmental resources by prehistoric land use on Poike peninsula, Easter Island (Rapa Nui). Rapa Nui J. 17, 34–41.

Mieth, A., Bork, H.R., 2005. History, origin and extent of soil erosion on Easter Island (Rapa Nui). Catena 63, 244–260.

Mieth, A., Bork, H.R., 2010. Humans, climate or introduced rats—which is to blame for the woodland destruction on prehistoric Rapa Nui (Easter Island)? J. Archaeol. Sci. 37, 417–426.

Mieth, A., Bork, H.R., 2012. Die Osterinsel—A Tour. Springer Spektrum, Berlin.

Mieth, A., Bork, H.R., 2015. Degradation of resources and successful land-use management on prehistoric Rapa Nui: two sides of the same coin. In: Cauwe, N., De Dapper, M. (Eds.), International Conference Easter Island: Collapse or Transformation? A State of the Art. Royal Academy of Overseas Science, Royal Museums of Art and History, Belgian Science Policy Office, Brussels, pp. 91–113.

Mieth, A., Bork, H.R., 2017. A vanished landscape—phenomena and eco-cultural consequences of extensive deforestation in the prehistory of Rapa Nui. In: Haoa Cardinali, S., Ingersoll, K.B., Stevenson, C.M. (Eds.), Cultural and Environmental Change on Rapa Nui. Routledge, London, pp. 32–58.

Mieth, A., Bork, H.R., Feeser, I., 2002. Prehistoric and recent land use effects on Poike peninsula, Easter Island (Rapa Nui). Rapa Nui J. 16, 89–95.

Montenegro, A., Avis, C., Weaver, A., 2008. Modeling the prehistoric arrival of the sweet potato in Polynesia. J. Archaeol. Sci. 35, 355–367.

Moreno-Mayar, J.V., Rasmussen, S., Seguin-Orlando, A., Rasmussen, M., Liang, M., Flåm, S.T., Lie, B.A., Gilfillan, G.D., Nielsen, S., Thorsby, E., Willerslev, E., Malaspinas, E.S., 2014. Genome-wide ancestry patterns in Rapanui suggest pre-European admixture with Native Americans. Curr. Biol. 24, 2518–2525.

Mucciarone, D.A., Dunbar, R.B., 2003. Stable iotope record of El Niño-Southern Oscillation events from Easter Island. In: Loret, J., Tanacredi, J.T. (Eds.), Easter Island. Scientific Exploration into the World's Environmental Problems in Microcosm. Kluwer Academic/Plenum, New York, pp. 113–132.

Mulloy, W., 1970a. A speculative reconstruction of techniques of carving, transporting and erecting Easter Island statues. Archaeol. Phys. Anthropol. Oceania 5, 1–23.

Mulloy, W., 1970b. Preliminary Report of the Restoration of Ahu Vai Uri, Easter Island. Bull. 2, International Fund for Monuments. Easter Island Committee, New York.

Mulloy, W., 1973. Preliminary Report of the Restoration of Ahu Uri a Urenga and Two Unnamed Ahu at Hanga Kio'e. Bull. 3, International Fund for Monuments. Easter Island Committee, New York.

Mulloy, W., 1974. Contemplate the Navel of the World. Américas 26, 25–33.

Mulloy, W., 1975. Investigation and Restoration of the Ceremonial Center of Orongo. Bull. 4, International Fund for Monuments. Easter Island Committee, New York.

Mulloy, W., 1979. A preliminary culture-historical research model for Easter Island. In: Echevarría, G., Arana, P. (Eds.), Las Islas Oceánicas de Chile. Universidad de Chile, Santiago de Chile, pp. 105–151.

Mulloy, W., 1997. Preliminary culture-historical research model for Easter Island. The Easter Island Bulletins of William Mulloy. World Monumets Fund & Easter Island Foundation, New York & Houston, pp. 97–111.

Mulloy, W., Figueroa, G., 1961. Cómo fue restaurado el Ahu Akivi en la Isla de Pascua. Bol. Universidad de Chile 27, 4–11.

Mulloy, W., Figueroa, G., 1978. The A-Kivi Vai-Teka complex and its relationship to Easter Island architectural prehistory. Asian and pacific Archaeology Series n° 8 University of Hawai'i, Manoa.

Mulrooney, M.A., 2013. An island-wide assessment of the chronology of settlement and land use on Rapa Nui (Easter Island) based on radiocarbon data. J. Archaeol. Sci. 40, 4377–4399.

Mulrooney, M., Ladefoged, T.N., Stevenson, C.M., Haoa, S., 2007. Empirical asessment of a pre-European societal collapse on Rapa Nui (Easter Island). In: Wallin, P., Martrinsson-Wallin, H. (Eds.), The Gotland Papers. Selected Papers from the VII International Conference on Easter Island and the Pacific. Migration, Identity, and Cultural Heritage. Gotland University Press, Sweden, pp. 141–153.

Mulrooney, M., Ladefoged, T.N., Stevenson, C.M., Haoa, S., 2009. The myth of A.D. 1680: new evidence from Hanga Ho'onu, Rapa Nui (Easter Island). Rapa Nui J. 23, 94–105.

Mulrooney, M.A., Bickler, S.H., Allen, M.S., Ladefoged, T.N., 2011. High-precision dating of colonization and settlement in East Polynesia. Proc. Natl. Acad. Sci. U. S. A. 108, E192–E194.

Muñoz-Rodríguez, P., Carruthers, T., Wood, J.R.I., Williams, B.R.M., Weitemier, K., Kronmiller, B., Ellis, D., Anglin, N.L., Longway, L., Harris, S.A., Rausher, M.D., Kelly, S., Liston, A., Scotland, R.W., 2018. Reconciling conflicting phylogenies in the origin of sweet potato and dispersal to Polynesia. Curr. Biol. 28, 1246–1256.

Negri, A.J., Adler, R.F., Shepherd, J.M., Huffman, G., Manyin, M., Neklin, E.J., 2004. A 16-year climatology of global rainfall from SSM/I highlighting morning versus evening effects. 13th American Meteorological Conference on Satellite Meteorology and Oceanography, Norfolk, P6.16.

Nunn, P.D., 2000. Environmental catastrophe in the Pacific Islands around A.D. 1300. Geoarchaeology 15, 715–740.

Nunn, P.D., 2007. Climate, Environment and Society in the Pacific During the Last Millennium. Elsevier, Amsterdam.

Nunn, P.D., Britton, J.M.R., 2001. Human-environment relationships in the Pacific Islands around A.D. 1300. Environ. Hist. 7, 3–22.

Nunn, P.D., Hunter-Anderson, R., Carson, M.T., Thomas, F., Ulm, S., Rowland, M.J., 2007. Times of plenty, times of less: last-millennium societal disruption in the Pacific basin. Hum. Ecol. 35, 385–401.

O'Connor, J.M., Stoffers, P., McWilliams, M.O., 1995. Time-space mapping of Easter Chain volcanism. Earth Planet. Sci. Lett. 136, 197–212.

O'Leary, M.H., 1988. Carbon isotopes in photosynthesis. Bioscience 38, 328–336.

Olsson, I.U., 1986. A study of errors in ^{14}C dates of peat and sediment. Radiocarbon 28, 429–435.

Orliac, C., 1990. *Sophora toromiro*, one of the raw materials used by Pascuan carvers: some examples in the collections of Musée de l'Homme. In: Esen-Baur, H.M. (Ed.), State and Perspectives of Scientific Research in Easter Island Culture, vol. 125. Courier Forschungsinstitut Senckenburg, Frankfurt, pp. 221–227.

Orliac, C., 2000. The woody vegetation of Easter Island between the early 14th and the mid-17th centuries AD. In: Stevenson, C.M., Ayres, W.S. (Eds.), Easter Island Archaeology. Research on Early Rapanui Culture. Easter Island Foundation, Los Osos, pp. 211–220.

Orliac, C., 2005. Research report the rongorongo tablets from easter island: botanical identification and ^{14}C dating. Archaeol. Ocean. 40, 115–119.

Orliac, C., 2007. Botanical identification of 200 Easter Island wood carvings. In: Wallin, P., Martrinsson-Wallin, H. (Eds.), The Gotland Papers. Selected Papers from the VII International Conference on Easter Island and the Pacific. Migration, Identity, and Cultural Heritage. Gotland University Press, Sweden, pp. 125–139.

Orliac, C., Orliac, M., 1996. Arbres et arbustes de l'Île de Pâques: composition et évolution de la flore depuis l'arrivée des Polynésiens. Comptes Rendues de la Mission Arqueologique l'Île de Pâques 1995. Ministère des Affaires Etrangères/CNRS/Consejo de Monumentos/CONAF.

Orliac, C., Orliac, M., 1998. The disappearance of Easter Island's forest: over-exploitation or climatic catastrophe? In: Stevenson, C.M., Lee, G., Morin, F.J. (Eds.), Easter Island in Pacific Context. Proceedings of the Fourth International Conference on Easter Island and East Polynesia. Easter Island Foundation, Los Osos, pp. 129–134.

Overpeck, J.T., 1996. Warm climatic surprises. Science 271, 1820–1821.

Overpeck, J., Whitlock, C., Huntley, B., 2003. Terrestrial biosphere dynamics in the climate system: past and future. In: Alverson, K.D., Bradley, R.S., Pedersen, T.F. (Eds.), Paleoclimate, Global Change and the Future. Springer, Berlin, pp. 81–103.

Owsley, D.W., Barca, K.G., Simon, V.E., Gill, G.W., 2016. Evidence for injuries and violent death. In: Stephan, V.H., Gill, G.W. (Eds.), Skeletal Biology of the Ancient Rapanui (Easter Islanders). Cambridge University Press, Cambridge, pp. 222–252.

Palmer, J.L., 1870. A visit to Easter Island, or Rapa Nui, in 1868. J. R. Geogr. Soc. 40, 167–181.

Parducci, L., Bennett, K.D., Ficetola, G.F., Alsos, I.G., Suyama, Y., Wood, J.R., Pedersen, M.W., 2017. The ancient plant DNA in lake sediments. New Phytol. 214, 924–942.

Pavel, P., 1990. Reconstruction of the transport of moai. In: Esen-Baur, H.M. (Ed.), State and Perspectives of Scientific Research in Easter Island Culture, vol. 125. Courier Forschungsinstitut Senckenburg, Frankfurt, pp. 141–144.

Peiser, B., 2005. From genocide to ecocide, the rape of Rapa Nui. Energy Environ. 16, 513–539.

Penk, A., Brückner, E., 1901–1909. Die Alpen im Eiszeitalter. Tauchnitz, Leipzig.

Peteet, D., Beck, W., Ortiz, J., O'Connell, S., Kurdyla, D., Mann, D., 2003. Rapid vegetational and sediment change from Rano Aroi crater, Easter Island. In: Loret, J., Tanacredi, J.T. (Eds.), Easter Island. Scientific Exploration into the World's Environmental Problems in Microcosm. Kluwer Academic/Plenum, New York, pp. 81–92.

Pietrusewsky, M., Ikehara-Quebral, R., 2001. Multivariate comparisons of Rapa Nui (Easter Island), Polynesian, and circum-Polynesian crania. In: Stevenson, C.M., Lee, G., Morin, F.J. (Eds.), Pacific 2000. Proceedings of the Fifth International Conference on Easter Island and the Pacific. Easter Island Foundation, Los Osos, pp. 457–494.

Pinart, A., 1878. Exploration de l'Île de Pâques. Bull. Soc. Géogr. 16, 193–213.

Piperno, D.R., 2006. Phytoliths. A Comprehensive Guide for Archaeologists and Paleoecologists. Altamira Press, Lanham.

Piperno, D.R., Holst, I., 1998. The presence of starch grains on prehistoric stone tools from the humid Neotropics: indications of early tuber use and agriculture in Panama. J. Archaeol. Sci. 25, 765–776.

Platt, J.R., 1964. Strong inference. Science 146, 347–353.

Polet, C., 2015. Starvation and cannibalsim on Easter Island? The contribution of the analysis of Rapanui human remains. In: Cauwe, N., De Dapper, M. (Eds.), International Conference Easter Island: Collapse or transformation? A State of the Art. Royal Academy of Overseas Science, Royal Museums of Art and History, Belgian Science Policy Office, Brussels, pp. 115–133.

Polet, C., Bocherens, H., 2016. New insights into the marine contribution to ancient Easter Islander's diet. J. Archaeol. Sci. Rep. 6, 709–719.

Pollard, J., Paterson, A., Welham, K., 2010. The Miro o'one: the archaeology of contact on Rapa Nui (Easter Island). World Archaeol. 42, 562–580.

Popper, K.R., 1959. The Logic of Scientific Discovery. Basic Books, New York.

Prebble, M., Dowe, J.L., 2008. The late Quaternary decline and extinction of palms on oceanic Pacific islands. Quat. Sci. Rev. 27, 2546–2567.

Prebble, M., Wilmshurst, J.M., 2009. Detecting the initial impact of humans and introduced species on island environments in Remote Oceania using paleoecology. Biol. Invasions 11, 1529–1556.

Puleston, C.O., Ladefoged, T.N., Haoa, S., Chadwick, O.A., Vitousek, P.M., Stevenson, C.M., 2017. Rain, sun, soil, and sweat: a consideration of population limits on Rapa Nui (Easter Island) beofre European contact. Front. Ecol. Evol. 5, 69.

Puleston, C.O., Ladefoged, T.N., Haoa, S., Chadwick, O.A., Vitousek, P.M., Stevenson, C.M., 2018. Response: commentary: rain, sun, soil, and sweat: a consideration of population limits on Rapa Nui (Easter Island) beofre European contact. Front. Ecol. Evol. 6, 72.

Püschel, T.A., Espejo, J., Sanzana, M.J., Benítez, H.A., 2014. Analising the folral elements of the lost tree of Easter Island: a morphometric comparison between the remaining ex-situ lines of the endemic extinct species *Sophora toromiro*. PLoS One 9, e115548.

Rainbird, P., 2002. A message for our future? The Rapa Nui (Easter Island) ecodisaster and Pacific environments. World Achaeol. 33, 436–451.

Rallu, J.L., 2007. Pre- and post-contact population in island Polynesia. In: Kirch, P.V., Rallu, J.L. (Eds.), Growth and Collapse of Pacific Island Societies. University of Hawai'i Press, Honolulu, pp. 13–34.

Rawlence, N.J., Lowe, D.J., Wood, J.R., Young, J.M., Churchman, G.J., Huang, Y.T., Cooper, A., 2014. Using palaeoenvironmental DNA to reconstruct past environments: progresses and prospects. J. Quat. Sci. 29, 610–626.

Ray, J.S., Mahoney, J.J., Duncan, R.A., Ray, J., Wessel, P., Naar, D.F., 2012. Chronology and geochemistry of lavas from the Nazca Ridge and Easter Seamount Chain: an ~30 Myr hotspot record. J. Petrol. 53, 1417–1448.

Raymo, M.E., 1994. The initiation of Northern Hemisphere Glaciation. Annu. Rev. Earth Planet. Sci. 22, 353–383.

Reimer, P.J., Baillie, M.G.L., Bard, E., Bayliss, A., Beck, J.W., Bertrand, C.J.H., Blackwell, P.G., Buck, C.E., Cutler, K.B., Damon, P.E., Edwards, R.L., Fairbanks, R.G., Friedrich, M., Guilderson, T.P., Hogg, A.G., Hughen, K.A., Kromer, B., McCormac, G., Manning, S., Ramsey, C.B., Reimer, R.W., Remmele, S., Southon, J.R., Stuiver, M., Talamo, S., Taylor, F.W., van der Plicht, J., Weyhenmeyer, C.E., 2004. IntCal04 terrestrial radiocarbon age calibration, 0–26 cal kyr BP. Radiocarbon 46, 1029–1058.

Reimer, P.J., Baillie, M.G.L., Bard, E., Bayliss, A., Beck, J.W., Blackwell, P.G., Ramsey, C.B., Buck, C.E., Burr, G.S., Edwards, R.L., Friedrich, M., Grootes, P.M., Guilderson, T.P., Haidas, I., Heaton, T.J., Hogg, A.G., Hughen, K.A., Kaiser, K.F., Kromer, B., McCormac, F.G., Manning, S.W., Reimer, R.W., Richards, D.A., Southon, J.R., Talamo, S., Turney, C.S.M., van del Plicht, J., WEyhenmeyer, C.E., 2009. Intcal09 and Marine09 radiocarbon age calibration curves, 0-50,000 years cal BP. Radiocarbon 51, 1111–1150.

Richards, R., 2008. Easter Island 1793-1861: Observations by Early Visitors Before the Slave Raids. Easter Island Foundation, Los Osos.

Richards, R., 2017. The impact of the whalers and other foreing visitors before 1862. In: Haoa Cardinali, S., Ingersoll, K.B., Stevenson, C.M. (Eds.), Cultural and Environmental Change on Rapa Nui. Routledge, London, pp. 179–187.

Rind, D., Peteet, D., 1985. Terrestrial conditions at the last glacial maximum and CLIMAP sea-surface temperature estimates: are they consistent? Quat. Res. 24, 1–22.

Roberts, N., 2014. The Holocene. An Environmental History. Wiley-Blackwell, Chichester.

Robinson, T., Stevenson, C.M., 2017. The cult of the Birdman: religious change at 'Orongo, Rapa Nui (Easter Island). J. Pacif. Archaeol. 8, 88–102.

Roman, S., Bullock, S., Brede, M., 2016. Coupled societies are more robust against collapse: a hypothetical look at Easter Island. Ecol. Econ. 132, 264–278.

Rorrer, K., 1998. Subsistence evidence from inland and coastal cave siteson Easter Island. In: Stevenson, C.M., Lee, G., Morin, F.J. (Eds.), Easter Island in Pacific Context. Proceedings of the Fourth International Conference on Easter Island and East Polynesia. Easter Island Foundation, Los Osos, pp. 193–198.

Roth, M., 1990. The conservation of the moai "Hanga Kio'e". Methods and consequences of the restoration. In: Esen-Baur, H.M. (Ed.), State and Perspectives of Scientific Research in Easter Island Culture, vol. 125. Courier Forschungsinstitut Senckenburg, Frankfurt, pp. 183–188.

Roullier, C., Benoit, L., McKey, D.B., Lebot, V., 2013. Historical collections reveal patterns of diffusion of sweet potato in Oceania obscured by modern plant movements and recombination. Proc. Natl. Acad. Sci. U. S. A. 110, 2205–2210.

Routledge, K., 1919. The Mystery of Easter Island. The Story of an Expedition. Shifton, Praed & Co., London.

Rull, V., 2008. Speciation timing and neotropical biodiversity: the Tertiary-Quaternary debate in the light of molecular phylogenetic evidence. Mol. Ecol. 17, 2722–2729.

Rull, V., 2010. Ecology and paleoecology: two approaches, one objective. Open Ecol. J. 3, 1–5.

Rull, V., 2011. Neotropical biodiversity: timing and potential drivers. Trends Ecol. Evol. 26, 508–513.

Rull, V., 2016a. The EIRA database: Last Glacial and Holocene radiocarbon ages from Easter Island's sedimentary records. Front. Ecol. Evol. 4, 44.

Rull, V., 2016b. Natural and anthropogenic drivers of cultural change at Easter Island: review and new insights. Quat. Sci. Rev. 150, 31–41.

Rull, V., 2018. Strong Fuzzy EHLFS: a general conceptual framework to address past records of environmental, ecological and cultural change. Quaternary 1, 10.

Rull, V., 2019. Human discovery and settlement of the remote Easter Island (SE Pacific). Quaternary 2, 15.

Rull, V., 2020a. The deforestation of easter Island. Biol. Rev. 95, 121–141.

Rull, V., 2020b. Drought, freshwater availability and cultural resilience on Easter Islad (SE Pacific) during the Little Ice Age. Holocene. 30, 774–780.

Rull, V., 2020c. Quaternary Ecology, Evolution and Biogeography. Academic Press, London.

Rull, V., Cañellas-Boltà, N., Sáez, A., Giralt, S., Pla, S., Margalef, O., 2010. Paleocology of Easter Island: evidence and uncertainties. Earth-Sci. Rev. 99, 50–60.

Rull, V., Cañellas-Boltà, N., Sáez, A., Margalef, O., Bao, R., Pla-Rabes, S., Valero-Garcés, B., Giralt, S., 2013. Challenging Easter Island's collapse: the need for interdisciplinary synergies. Front. Ecol. Evol. 1, 3.

Rull, V., Cañellas-Boltà, N., Margalef, O., Sáez, A., Pla-Rabes, S., Giralt, S., 2015. Late Holocene vegetation dynamics and deforestation in Rano Aroi: implications for Easter Island's ecological and cultural history. Quat. Sci. Rev. 126, 219–226.

Rull, V., Cañellas-Boltà, N., Margalef, O., Pla-Rabes, S., Sáez, A., Giralt, S., 2016. Three millennia of climatic, ecological and cultural change on Easter Island: a synthetic overview. Front. Ecol. Evol. 4, 29.

Rull, V., Lara, A., Rubio-Inglés, M.J., Giralt, S., Gonçalves, V., Raposeiro, P., Hernández, A., Sánchez-López, G., Vázquez-Loureiro, D., Bao, R., Masqué, P., Sáez, A., 2017. Vegetation and landscape dynamics under natural and anthropogenic forcing on the Azores Islands: a 700-year pollen record from the São Miguel Island. Quat. Sci. Rev. 159, 155–168.

Rull, V., Montoya, E., Seco, I., Cañellas-Boltà, N., Giralt, S., Margalef, O., Pla-Rabes, S., D'Andrea, W., Bradley, R., Sáez, A., 2018. CLAFS, a holistic climatic-ecological anthropogenic hypothesis on Easter Island's deforestation and cultural change: proposals and testing prospects. Front. Ecol. Evol. 6, 32.

Rutherford, S., Shepardson, B., Stephen, J., 2008. A premiminary lichenometry study on Rapa Nui—the Rapa Nui Youth Involvement Program Report. Rapa Nui J. 22, 40–47.

Sáez, A., Valero-Garcés, B., Giralt, S., Moreno, A., Bao, R., Pueyo, J.J., Hernández, A., Casas, D., 2009. Glacial to Holocene climate changes in the SE Pacific. The Raraku Lake sedimentary record (Easter Island, 27°S). Quat. Sci. Rev. 28, 2743–2759.

Sawada, M., Koezuka, T., Kohdzuma, Y., Inoue, S., Bahamóndez, M., 2001. In-situ weathering tests of conservation materials applied to volcanic tuff samples from Ahu Tomgariki, Easter Island. In: Stevenson, C.M., Lee, G., Morin, F.J. (Eds.), Pacific 2000. Proceedings of the Fifth International Conference on Easter Island and the Pacific. Easter Island Foundation, Los Osos, pp. 525–532.

Schumacher, Z., 2013. A Geo-Spatial Database of the Monumental Sanctuary (moai) of Easter Island, Chile. (MA dissertation). California State University, Long Beach.

Seco, I., Rull, V., Montoya, E., Cañellas-Boltà, N., Giralt, S., Margalef, O., Pla-Rabes, S., D'Andrea, W.J., Bradley, R.S., Sáez, A., 2019. A continuous palynological record of forest clearing at Rano Kao (Easter Island, SE Pacific) during the last millennium: preliminary report. Quaternary 2, 22.

Selling, O.H., 1946–1948. Studies in Hawaiian Pollen Statistics, vols. I–III. Bernice P. Bishop Museum Spec. Publ, Honolulu.

Shaw, L.C., 1998. Landscape and the meaning of place in Easter Island burial practices. In: Stevenson, C.M., Lee, G., Morin, F.J. (Eds.), Easter Island in Pacific Context. Proceedings of the Fourth International Conference on Easter Island and East Polynesia. Easter Island Foundation, Los Osos, pp. 218–222.

Shepardson, B., 2006. Explaining Spatial and Temporal Patterns of Energy Investment in the Prehistory Sanctuary of Rapa Nui (Easter Island). (Ph.D. dissertation). University of Hawai'i, Manoa.

Shepardson, B., 2009. Moai of Rapa Nui. Designing, Coding and Administering an Online Archaeological Data Warehouse Using GIS, Web, and Spreadsheet Technologies. Terevaka.net Data Community.

Shepherd, L.D., Thiedemann, M., Lehnebach, C., 2020. Genetic identification of historic *Sophora* (Fabaceae) specimens suggests toromiro (*S. toromiro*) from Rapa Nui/Easter Island may have been in cultivation in Europe in the 1700s. N. Z. J. Bot., https://doi.org/10.1080/0028825X.2020.1725069.

Sherwood, S.C., Van Tilburg, J.A., Barrier, C.R., Horrocks, M., Dunn, R.K., Ramírez-Aliaga, J.M., 2019. New excavations in Easter island's statue quarry: soil fertility, site formation and chronology. J. Archaeol. Sci. 111, 104994.

Simpson, D.F., Dussubieux, L., 2018. A collapse narrative? Geochemistry and spatial distribution of basalt quarries and fine-grained artifacts reveal communal use of stone on Rapa Nui (Easter Island). J. Archaeol. Sci. Rep. 18, 370–385.

Skottsberg, C., 1920–1956. The Natural History of Juan Fernandez and Easter Island. 6 vols. Almqvist & Wiksells, Uppsala.

Smith, C.S., 1961. A temporal sequence derived from certain ahu. In: Heyerdahl, T., Ferdon, E. (Eds.), Reports of the Norwegian Archaeological Expedition to Easter Island and the East Pacific. Archaeology of Easter Island, vol. 1. Allen & Unwin, London, pp. 181–218.

Smith, C.S., 2000. The archaeological investigations on Easter Island in historical perspective. In: Stevenson, C.P., Ayres, W.S. (Eds.), Easter Island Archaeology: Research on Early Rapanui Culture. Easter Island Foundation, Los Osos, pp. 3–6.

Smith, A.D., 2003. A reappraisal of stress field and connective roll models for the origin and distribution of Cretaceous to recent intraplate volcanism in the Pacific Basin. Int. Geol. Rev. 45, 287–302.

Smol, J.P., Birks, H.J.B., Last, W.M. (Eds.), 2001a. Tracking Environmental Change Using Lake Sediments. Terrestrial, Algal, and Siliceous Indicators, vol. 3. Kluwer, Dordrecht.

Smol, J.P., Birks, H.J.B., Last, W.M. (Eds.), 2001b. Tracking Environmental Change Using Lake Sediments. Zoological Indicators, vol. 4. Kluwer, Dordrecht.

Steadman, D.W., Vargas, P., Cristino, C., 1994. Stratigraphy, chronology, and cultural context of an early faunal assemblage from Easter Island. Asian Perspect. 33, 79–96.

Stefan, V.H., 2001. Origin and evolution of the Rapanui of Easter Island. In: Stevenson, C.M., Lee, G., Morin, F.J. (Eds.), Pacific 2000. Proceedings of the Fifth International Conference on Easter Island and the Pacific. Easter Island Foundation, Los Osos, pp. 495–522.

Stenseth, N.C., Voje, K.L., 2009. Easter Island: climate change might have contributed to past cultural and societal changes. Clim. Res. 39, 111–114.

Stevenson, C.M., 1986. The sociopolitical structure of the southern coastal area of Easter Island: AD 1300-1864. In: Kirch, P.V. (Ed.), Island Societies: Archaeological Approaches to Evolution and Transformation. Cambridge University Press, Cambridge, pp. 69–77.

Stevenson, C.M., 2002. Territorial divisions on Easter Island in the 16th century: evidence from the distribution of ceremonial architecture. In: Ladefoged, T.N., Graves, M. (Eds.), Pacific Landscapes: Archaeological Approaches. Easter Island Foundation, Los Osos, pp. 221–230.

Stevenson, C.M., Novak, S.W., 1988. Obsidian hydration dating by infrared spectroscopy: method and calibration. J. Archaeol. Sci. 38, 1716–1726.

Stevenson, C.M., Williams, C., 2018. The temporal occurrence and possible uses of obsidian *mata'a* on Rapa Nui (Easter Island, Chile). Archaeol. Ocean. 53, 92–102.

Stevenson, C.M., Wozniak, J., Haoa, S., 1999. Prehistoric agricultural production on Easter Island (Rapa Nui), Chile. Antiquity 73, 801–812.

Stevenson, C.M., Ramírez, J.M., Haoa, S., Allen, T., 2000. Archaeological investigations at 'Anakena beach and other near-coastal locations. In: Stevenson, C.P., Ayres, W.S. (Eds.), Easter Island Archaeology: Research on Early Rapanui Culture. Easter Island Foundation, Los Osos, pp. 147–172.

Stevenson, C.M., Ladefoged, T.N., Novak, S.W., 2013. Prehistoric settlement chronology on Rapa Nui, Chile: obsidian hydration dating using infrared photoacoustic spectroscopy. J. Archaeol. Sci. 40, 3021–3030.

Stevenson, C.M., Puleston, C.O., Vitousek, P.M., Chadwick, O.A., Haoa, S., Ladefoged, T.N., 2015. Variation in Rapa Nui (Easter Island) land use indicates production and population peaks prior to European contact. Proc. Natl. Acad. Sci. U. S. A. 112, 1025–1030.

Storey, A.A., Ramírez, J.M., Quiroz, D., Burley, D.V., Addison, D.J., Walter, R., Anderson, A.J., Hunt, T.L., Athens, J.S., Huynen, L., Matisoo-Smith, E.A., 2007. Radiocarbon and DNA evidence for pre-Columbian introduction of Polynesian chickens to Chile. Proc. Natl. Acad. Sci. U. S. A. 104, 10335–10339.

Stuiver, M., Reimer, P.J., 1986. Computer program for radiocarbon age calibration. Radiocarbon 28, 1022–1030.

Stuiver, M., Reimer, P.J., Bard, E., Beck, J.W., Burr, G.S., Hughen, K.A., Kromer, B., McCormack, F.G., van der Plicht, J., Spurk, M., 1998. INTCAL98 radiocarbon age calibration, 24,000-0 cal BP. Radiocarbon 40, 1041–1083.

Sugita, S., 2007a. Theory of quantitative reconstruction of vegetation. I: pollen from large sites REVEALS regional vegetation composition. Holocene 17, 229–241.

Sugita, S., 2007b. Theory of quantitative reconstruction of vegetation. II: all you need is LOVE. Holocene 17, 243–257.

Thompson, W.J., 1891. The Pito te Henua, or Easter Island. Rep. US Nat. Mus, Washington, pp. 447–552.

Thompson, V.A., Lebrasseur, O., Austin, J.J., Hunt, T.L., Burney, D.A., Denham, T., Rawlence, N.J., Wood, J.R., Gongora, J., Flink, L.G., Linderholm, A., Dobney, K., Larson, G., Cooper, A., 2014. Using ancient DNA to study the origins and dispersal of ancestral Polynesian chickens across the Pacific. Proc. Natl. Acad. Sci. U. S. A. 111, 4826–4831.

Thorsby, E., 2007. Evidence of an early Amerindian contribution to the Polynesian gene pool on Easter Island. In: Wallin, P., Martrinsson-Wallin, H. (Eds.), The Gotland Papers. Selected Papers from the VII International Conference on Easter Island and the Pacific. Migration, Identity, and Cultural Heritage. Gotland University Press, Sweden, pp. 285–295.

Thorsby, E., 2016. Genetic evidence of a contribution of Native Americans to the early settlement of Rapa Nui (Easter Island). Front. Ecol. Evol. 4, 118.

Torres Hochstetter, F.T., Rapu, S., Lipo, C.P., Hunt, T.L., 2011. A public database of archaeological resources on Easter Island (Rapa Nui) using Google Earth. Lat. Am. Antiq. 22, 385–397.

Trenberth, K.E., 2019. El Niño Sothern Oscillation (ENSO). Encycl. Ocean Sci. 6, 420–432.

Tromp, M., Dudgeon, J.V., 2015. Differentiating dietary and non-dietary microfossils extracted from human dental calculus: the importance of sweet potato to ancient diet on Rapa Nui. J. Archaeol. Sci. 54, 54–63.

Van Balgooy, M.M.J., 1960. Preliminary plant geographical analysis of the Pacific as based on the distribution of phanerogam genera. Blumea 10, 385–430.

Van der Hammen, T., 1974. The Pleistocene changes of vegetation and climate in tropical South America. J. Biogeogr. 1, 3–26.

Van der Leeuw, S., Costanza, R., Aulenbach, S., Brewer, S., Cornell, S., Crumley, C., Dearing, J.A., Downy, C., Graumlich, L.J., Heckbert, S., Hegmon, M., Hibbard, K., Jackson, S.T., Kubiszewski, I., Sinclair, P., Sörlin, S., Steffen, W., 2011. Toward an integrated history to guide the future. Ecol. Soc. 16, 2.

Van Geel, B., Buurman, J., Brinkkemper, O., Schelvis, J., Aptroot, A., van Reenen, G., Hakbijl, T., 2003. Environmetal reconstruction of a Roman Period settlement site in Uitgeest (The Netherlands), with special reference to coprophilous fungi. J. Archaeol. Sci. 30, 873–883.

Van Tilburg, J.A., 1994. Easter Island: Archaeology, Ecology, and Culture. Smithsonian Inst. Press, Washington.

Van Tilburg, J.A., Ralston, T., 2005. Megaliths and mariners: experimental archaeology on Easter Island. In: Johnson, K. (Ed.), Onward and Upward! Papers in Honor of Clement W. Meighan, pp. 279–303.

Vandergoes, M.J., Prior, C.A., 2003. AMS dating of pollen concentrates—a methodological study of late Quaternary sediments from South Westland, New Zealand. Radiocarbon 45, 479–491.

Vannière, B., Power, M.J., Roberts, N., 2011. Circum-Mediterranean fire activity and climate changes during the mid-Holocene environmental transition (8500-2500 cal. BP). Holocene 21, 53–73.

Vargas, P., Cristino, C., Izaurieta, R., 2006. 1.000 años en Rapa Nui. Cronología del Asentamiento. Editorial Universitaria, Santiago de Chile.

Vegas-Vilarrúbia, T., Rull, V., Montoya, E., Safont, E., 2011. Quaternary palaeoecology and nature conservation: a general review with example from the Neotropics. Quat. Sci. Rev. 30, 2361–2388.

Vezzoli, L., Acocella, V., 2009. Easter Island, SE Pacific: and end-member type of hotspot volcanism. Geol. Soc. Am. Bull. 121, 869–886.

Volterra, V., 1926. Variazioni e fluttuazioni del numero d'individui in specie animale conviventi. Mem. R. Accad. Naz. Lincei 2, 31–113.

Von Däniken, E., 1980. Chariots of the Gods. Berkley Books, New York.

Walker, M., Gibbard, P., Head, M.J., Berkelhammer, M., Björk, S., Cheng, H., Cwynar, L., Fischer, D., Gkinis, V., Long, A., Lowe, J., Newham, R., Rasmussen, S.O., Weiss, H., 2019. Formal subdivision of the Holocene Series/Epoch: a summary. J. Geol. Soc. India 93, 135–141.

Wallin, P., Martisson-Wallin, H., Possnert, G., 2010. Re-dating Ahu Nau Nau and the settlement at 'Anakena, Rapa Nui. In: Wallin, P., Martinsson-Wallin, H. (Eds.), The Gotland Papers. Selected Papers from the VII International Conference on Easter Island and the Pacific. Gotland University, Sweden, pp. 37–46.

Weisler, M.I., Green, R.C., 2011. Rethinking the chronology of colonization of Southeast Polynesia. In: Jones, T.L., Storey, A.A., Matisoo-Smith, E.A., Ramírez-Aliaga, J.M. (Eds.), Pre-Columbian Contacts with the New World. Altamira Press, Lanham, pp. 223–246.

West, D.C., Shugart, D.B., Botkin, D.B., 1981. Forest Succession, Concepts and Applications. Springer, New York.

West, K., Collins, C., Kardailsky, O., Kahn, J., Hunt, T.L., Burley, D.V., Matisoo-Smith, E., 2017. The Pacific rat race to Easter Island: tracking the prehistoric dispersal of *Rattus exulans* using ancient mitochondrial genomes. Front. Ecol. Evol. 5, 52.

Williams, J.W., Jackson, S.T., 2007. Novel climates, no-analog communities, and ecological surprises. Front. Ecol. Environ. 5, 475–482.

Willis, K.J., Bailey, R.M., Bhagwat, S.A., Birks, H.J.B., 2010. Biodiversity baselines, thresholds and resilience: testing predictions and assumptions using palaeoecological data. Trends Ecol. Evol. 25, 583–591.

Wilmshurst, J.M., Hunt, T.L., Lipo, C.P., Anderson, A.J., 2011. High-precision radiocarbon dating shows recent and rapid initial human colonization of East Polynesia. Proc. Natl. Acad. Sci. U. S. A. 108, 1815–1820.

Wozniak, J.A., 1999. Prehistoric horticultural practices on Easter Island: lithic mulched gardens and field streams. Rapa Nui J. 13, 95–99.

Wozniak, J.A., 2001. Landscapes of food production on Easter Island: successful subsistence strategies. In: Stevenson, C.M., Lee, G., Morin, F.J. (Eds.), Pacific 2000. Proceedings of the Fifth International Conference on Easter Island and the Pacific. Easter Island Foundation, Los Osos, pp. 91–101.

Wozniak, J.A., 2017. Subsistence strategies on Rapa Nui (Easter Island). Prehistoric gardening practices on Rapa Nui and how they related to current farming practices. In: Haoa Cardinali, S., Ingersoll, K.B., Stevenson, C.M. (Eds.), Cultural and Environmental Change on Rapa Nui. Routledge, London, pp. 87–112.

Wozniak, J.A., Horrocks, M., Cummings, L., 2007. Plant microfossil analysis of deopists from Te Niu, Rapa Nui, demonstrates forest disruption c. AD 1300 and subsequent dryland multi-cropping. In: Wallin, P., Martrinsson-Wallin, H. (Eds.), The Gotland Papers. Selected Papers from the VII International Conference on Easter Island and the Pacific. Migration, Identity, and Cultural Heritage. Gotland University Press, Sweden, pp. 111–124.

Wright, H.E., 1977. Quaternary vegetation history—some comparisons between Europe and America. Annu. Rev. Earth Planet. Sci. 5, 123–158.

Yen, D.E., 1974. The Sweet Potato in Oceania: An Essay in Ethnobotany. Bishop Museum Press, Honolulu.

Zalasiewicz, J., Williams, M., Heywood, A., Ellis, M., 2011. The Anthropocene: a new epoch og geological time? Philos. Trans. R. Soc. A 369, 835–841.

Zizka, G., 1991. Flowering Plants of Easter Island. Palmarum Hortus Francofurtensis, Frankfurt Main.

Species index

Note: Page numbers followed by *f* indicate figures and *t* indicate tables.

A

Acacia spp., 25
Acalypha sp., 25, 160–164*t*
Achnanthes cf. *abundans*, 168–170
Adianthum cf. *raddianum*, 14–16
Ageratum conyzoides, 160–164*t*
Agrostis avenacea, 17–18*t*
Albizia lebbeck, 39
Alona weinecki, 168–170
Alphitonia cf. *zizyphoides*, 176*t*
Anisothecium hoockeri, 14*t*
Anredera cordifolia, 25
Apium, 139–141, 210–212
    A. amni, 160–164*t*
    A. prostratum, 17–18*t*, 39, 160–164*t*
Araucaria, 14–16
Asplenium, 99
    A. adiantoides, 160–164*t*
    A. obtusatum, 160–164*t*
    A. obtusatum var. *obtusatum*, 15*t*
    A. polyodon, 39
    A. polyodon var. *squamulosum*, 15*t*
Australopithecus, 235
Axonopus
    A. compressus, 16
    A. paschalis, 16, 17–18*t*, 21–23

B

Bidens Pilosa, 160–164*t*
Blechnum pascale, 15*t*
Blindia magellanica, 14*t*
Brachymenium indicum, 14*t*
Bromus catharticus, 17–18*t*
Brousonetia papyrifera, 201–203
Broussonetia, 174, 177, 201
    B. papyrifera, 21–23, 160–164*t*, 176*t*,
    186–189, 192, 226*t*
Bryum sp., 14*t*
    B. argenteum, 14*t*

C

Caesalpinia major, 17–18*t*, 25, 176*t*
Calystegia sepium, 17–18*t*

Campylopus
    C. clavatus, 14*t*
    C. introflexus, 14*t*
    C. vesticaulis, 14*t*
Campylotheca sp., 160–164*t*
Canavalia sp., 160–164*t*
Capparis sp., 160–164*t*
Casuarina, 139–141
Casuarina equisetifoila, 39, 160–164*t*
Catharanthus roseus, 25
Cedrus, 14–16
Centaurea cyanus, 160–164*t*
Centaurium spicatum, 17–18*t*
Cephaloziella sp., 13*t*
Cerastium glomeratum, 160–164*t*
Ceratodon purpureus, 14*t*
Chenia leptophyla, 14*t*
Chenopodium glaucum, 17–18*t*, 39
Chrysogonum sp., 160–164*t*
Cocos, 135–136
    C. nucifera, 19–21, 23–24, 24*f*, 39,
    135–136
Colocasia, 174
    C. esculenta, 21–23, 191, 197–199,
    201–203, 226*t*
Cololejeunea minutissima ssp. *myriocarpa*, 13*t*
Conyza bonariensis, 160–164*t*
Coprosma sp., 23, 138–139, 143–145, 149,
    157–158, 160–164*t*, 176*t*, 177
Cordyline, 174
    C. fruticosa, 21–23
Cotula australis, 160–164*t*
Crisium vulgare, 160–164*t*
Crotalaria spp., 19–21
    C. grahamiana, 25
Cupressus, 14–16
Curcuma, 174
    C. longa, 21–23
Cyathea sp., 160–164*t*
Cyclosorus, 99
    C. cf. *parasiticus*, 14–16
    C. interruptus, 39
Cynara scolymus, 160–164*t*

Cynodon dactylon, 17–18*t*
Cyperus, 139
   C. cyperoides, 17–18*t*, 160–164*t*
   C. eragrostis, 17–18*t*
   C. polystachius, 160–164*t*
   C. vegetus, 160–164*t*
Cyrtomium falcatum, 14–16

D

Danthonia paschalis, 16, 17–18*t*, 25
Davallia solida, 15*t*, 160–164*t*
Dianella, 23
Dianthus caryophyllus, 160–164*t*
Dichelachne
   D. crinita, 17–18*t*
   D. micrantha, 17–18*t*
Dicranella
   D. cardotii, 14*t*
   D. hawaiica, 14*t*
Dioscorea alata, 21–23, 191, 226*t*
Diplazium fuenzalidae, 15*t*
Diploschistes anactinus, 12
Ditrichum dificile, 14*t*
Dodonaea viscosa, 19–21, 25, 39
Doodia paschalis, 160–164*t*
Dryopteris
   D. gongyloides, 160–164*t*
   D. karwinskyana, 15*t*
   D. parasitica, 160–164*t*
   D. spinosa, 160–164*t*
Dumortiera hirsuta, 13*t*

E

Elaeocarpus rarotongensis, 176*t*
Elaphoglossum skottsbergii, 15*t*, 160–164*t*
Ephedra spp., 160–164*t*
Ephorbia
   E. hirta, 160–164*t*
   E. peplus, 160–164*t*
   E. serpens, 160–164*t*
Erigeron linifolius, 160–164*t*
Eucalyptus spp., 19–21, 24*f*, 25, 99, 143–145
Euphorbia, 139
   E. serpens, 17–18*t*, 148

F

Fabronia jamesonii, 14*t*
Ficus carica, 160–164*t*

Fissidens
   F. pascuanus, 14*t*
   F. pellucidus, 14*t*
Foeniculum vulgare, 160–164*t*
Frullania ericoides, 13*t*

G

Galinsoga parviflora, 160–164*t*
Grevillea robusta, 25

H

Haplopteris ensiformis, 15*t*
Hibiscus tiliaceous, 25
Homo, 235
Hypochoeris radicata, 160–164*t*

I

Ipomoea, 174
   I. batatas, 21–23, 22*f*, 51–52, 55–57, 56*f*,
      191, 197–199, 201–203, 226*t*
   I. pes-caprae, 17–18*t*, 201–203
Isopterygium albescens, 14*t*

J

Jackiella javanica, 13*t*
Juania australis, 25–26
Jubaea, 25–26
   J. chilensis, 25–26, 26*f*, 135–136, 160–164*t*

K

Kyllinga brevifolia, 17–18*t*, 19–21, 160–164*t*

L

Lactuca sativa, 160–164*t*
Lagenaria
   L. siceraria, 21–23, 22*f*, 191, 226*t*, 229–230
Lantana camara, 25
Larix, 14–16
Lecidea paschalis, 12
Lejeunea
   L. flava, 13*t*
   L. minutiloba, 13*t*
Leptobryum pyriforme, 14*t*
Lophocolea aberrans, 13*t*
Lycium
   L. carolinianum, 17–18*t*, 160–164*t*
   L. sandwicense, 39
Lycopodium sp., 160–164*t*

M

Macaranga sp., 23, 138–139, 160–164*t*
Macromitrium sp., 14*t*
Marchantia bertoana, 13*t*
Marsupidium knightii, 13*t*
Melia, 139–141
  M. azedarach, 19–21, 25, 160–164*t*
Melinis spp., 25
  M. minutiflora, 39
Metrosideros sp., 23, 160–164*t*
Meyenia sp., 168–170
Microlepia strigosa, 15*t*, 160–164*t*
Microsorum parskii, 15*t*, 39
Morus sp., 160–164*t*
Musa sp., 21–23, 174, 192, 194–196,
  201–203, 214, 226*t*
Myrsine, 176*t*

N

Navicula goeppertiana var. *monita*, 168–170
Nephrolepis cf. *cordifolia*, 14–16
Nicotiana tabacum, 25
Nitschia cf. *vidovichii*, 168–170

O

Onychoprion fuscatus, 67–68
Ophiglossum
  O. lusitanicum subsp. *coriaceum*, 15*t*
Ophioglossum
  O. coriaceum, 160–164*t*
  O. reticulatum, 15*t*

P

Papillaria crocea, 14*t*
Paronychia sp., 160–164*t*
Paschalococos, 177
  P. disperta, 17–18*t*, 25–26, 135–136, 176*t*
Paspalum
  P. forsterianum, 16, 17–18*t*, 39
  P. scrobiculatum, 19–21
Pennisetum clandestinum, 25
Philonotis astata, 14*t*
Physcia ahu, 12
Pinnularia latevittata, 168–170
Pinus, 14–16, 201
Pittosporum, 176*t*
Plantago, 139, 143–145
  P. lanceolata, 160–164*t*

Pneumatopteris costata var. *hispida*, 15*t*
Pohlia sp., 14*t*
Polycarpon tetraphyllum, 160–164*t*
Polygonum, 99, 139–141, 181–183, 201,
  208–210
  P. acuminatum, 17–18*t*, 25, 99–100,
  138–139, 160–164*t*
Polypodium scolopendria, 160–164*t*
Polystichum fuentesii, 15*t*
Portulaca oleracea, 17–18*t*
Potamogeton sp., 23, 160–164*t*
Premna cf. *serratifolia*, 176*t*
Pritchardia, 134–136
Psidium, 139–141
  P. guajava, 19–21, 25, 160–164*t*
Psilotum nudum, 15*t*, 160–164*t*
Psychotria, 176*t*
Psydrax cf. *odorata*, 176*t*
Pteris sp., 160–164*t*
Ptychomitrium subcylindricum, 14*t*
Pycreus polystachyos, 17–18*t*, 19–21, 99–100
Pyrrhobryum spiniforme, 14*t*

R

Racopilum cuspidigerum, 14*t*
Rattus concolor, 135–136
Riccardia tenerrima, 13*t*
Robinia pseudoacacia, 25
Rumex, 139
  R. obtusifolius, 160–164*t*
Rytidosperma paschale, 16

S

Saccharum officinarum, 21–23
Samolus repens, 17–18*t*
Sapindus, 138–139, 177
  S. saponaria, 39, 160–164*t*, 176*t*
Sarscypridopsis cf. *elisabethae*, 168–170
Scirpus, 111–112, 138–141, 143–145, 148,
  170–171, 183–189, 187–188*f*, 197*f*,
  199, 201
  S. californicus, 17–18*t*, 19–21, 25, 99–101,
  138–139, 160–164*t*, 170–171, 173*f*,
  196–197
Secale cereale, 49–50
Sematophyllum
  S. aberrans, 14*t*
  S. brachycladulum, 14*t*

Setaria sphacelata, 25
Sisyrinchium, 23, 200*f*, 201, 216, 217*t*
  S. michrantum, 201
Solanum, 148, 181–183
  S. forsteri, 17–18*t*, 160–164*t*
Sonchus
  S. asper, 160–164*t*
  S. oleraceous, 160–164*t*
Sophora, 138–141, 177, 197–199, 208–210
  S. cassioides, 39–40
  S. toromiro, 16, 17–18*t*, 19*f*, 25, 30–33,
    39–40, 78*f*, 149, 160–164*t*, 176*t*
Sporobolus africanus, 17–18*t*, 19–21
Sporormiella, 49–50, 210–212, 223–225, 241*f*
Syzygium cf. *malacense*, 176*t*

T
Taraxacum officinale, 160–164*t*
Tetragonia tetragonoides, 17–18*t*, 39
Thelypteris interrupta, 15*t*
Thespesia, 177
  T. populnea, 19–23, 30–33, 39, 176*t*

Tortella humilis, 14*t*
Trema sp., 23, 138–141, 160–164*t*
Trematodon pascuanus, 14*t*
Triumfetta, 138–141, 148, 174, 177, 194–196
  T. semitriloba, 17–18*t*, 25, 39, 149,
    160–164*t*, 176*t*
Typha, 23
  T. angustifolia, 160–164*t*

V
Verbena, 205–208, 214–216
  V. litoralis, 23, 205, 207*f*, 214–215, 217*t*
Vittaria, 99
  V. elongata, 160–164*t*
  V. ensiformis, 39

W
Weissia controversa, 14*t*

X
Xylosma, 176*t*

Subject index

Note: Page numbers followed by *f* indicate figures and *t* indicate tables.

A

AA code, 114–115
Accelerator mass spectrometry (AMS) radiocarbon dating, 107–109
AD 1300 event, 95–97
Age-class frequency histograms, 107–109, 108*f*
Age-depth model, 167–168, 169*f*, 180–181
  Rano Aroi, 142–143, 142*f*
  Rano Kao, 139–141, 140*f*, 184–186, 185*f*, 187*f*
  Rano Raraku, 111–112, 111*f*, 137–138, 137*f*, 180–181, 181*f*
Ahu Moai phase, 44–46
Ahu Nau Nau, 63–64, 63*f*
Ahu Tongariki, 36–37
  moai complex of, 28*f*
Amerindians, arrival of, 168–170, 215–216
Ancient Rapanui society, 60
  Birdman cult, 67–72
  collapse/resilience, 74–80
  deforestation, 72–74
  demography, 80–82, 81*t*
  genocide, 83–85
  land distribution, 60–62, 61*f*
  moai cult, 63–67
  sociopolitical organization, 60–62
Angiosperms, 12
  features of, 16, 17–18*t*
Anthropochore, 16, 19–21, 214–215
Anthropogenic dispersal, of sweet potato, 55–57, 58*f*
Anthropogenic transformations, 237
Aquatic vegetation
  in Rano Aroi, 99
  in Rano Raraku, 100–101
Archaeological colonization, 49, 59–60
Archaeological heritage, 27–33
  conservation of, 36–38
  restoration of, 44
Archives, paleoecological, 112–113
Ariki Mau, 62
Azores Islands, 49–50, 59–60

B

Back-and-forth hypothesis, 55–59, 56*f*
Bacon age-depth model, 200*f*, 201
Bacon Bayesian approach, 199
Before Common Era (BCE), 42
Beta code, 114–115
Birdman cult, 46–47, 60, 62, 67–72, 228
  manutara (sooty tern) and, 67–68, 69*f*
  petroglyphs in, 68–69, 70*f*
  radiocarbon and obsidian hydration dates, 68–69, 71*f*
Bryophytes, 12–13

C

CALIB 7.1, 107–109
Calibrated (2σ) radiocarbon dates, on totora seeds/fruits, 192–194, 194*f*
Center for Acceleration Mass Spectrometry (CAMS) code, 114–115
Ceremonial village, of Orongo, 68–69, 70*f*
Charcoal content, in Rano Raraku, 172–173, 173*f*
Chiefdom Integration, 44–46
Chronology, prehistoric, 44–48
Clans, 60–62, 61*f*
Climate, 5–11
Climate changes, 156–157, 218–221
Climatic hydrological balance, 5–6
Climatic shifts, 218
Collapse/resilience, ancient Rapanui society, 74–80
Collapse theory, 165
Colonization, 49
  archaeological, 49, 59–60
Common Era (CE), 42
CONAF. *See* National Forestry Commission (CONAF)
Conservation
  of archaeological heritage, 36–38
  of native and Polynesian-introduced flora, 39–40
Coprophilous fungi, 49–50, 210–212
Corals, distribution of, 112, 113*f*

Coring intensification and reanalysis
  Rano Kao, 184–189, 185*f*, 187*f*
  Rano Raraku, 180–184, 181–182*f*, 184*f*
  soil investigations, 188*f*, 189–191, 190*f*

D

Dark Era, 42–44
Database of Easter Island Archaeology, 33
Decadence Period/Huri Moai Period, 44–46
Deforestation, 72–74, 221–225, 242–244
  chronology, 155–156
  continuous records, 203–212
Demography, 80–82, 81*t*
Diffusal Spectral Reflectance (DSR), 174
Discovery, 49–50
Dual determinism, 239–240

E

Early Period/Ahu Moroki, 44–46
Easter Island, 12
  archaeological heritage, 27–33
  bryophytes in, 12–13
  climate of, 5–11, 6–7*f*, 149
  conservation, 33–40
    of archaeological heritage, 36–38
    of native and Polynesian-introduced
      flora, 39–40
  Environmental-Human-Landscape
    Feedbacks and Synergies (EHLFS)
    approach in, 239–244, 240–242*f*
  freshwater sources of, 8–11, 9*f*
  geography and geology, 1–5
  geologic map, 3–5, 5*f*
  human and climatic impact, 158–159
  human-deterministic approaches,
    239–240, 240*f*
  human presence on, 217–218, 217*t*
  hydrology, 5–11, 10*f*
  hypothetical cause-effect model for
    ecological events, 147–148, 148*f*
  landscape and demographic inferences,
    149–152, 153*f*
  landscape deterioration, 25–27
  land use, 12
  lichens in, 12–13
  location of, 1–2
  paleoclimatic and paleoecological
    context, 240, 241*f*

paleoecology on, 93–94
  prehistoric paleoecological studies
    climate change, 218–221, 219*f*
    cultural consequences, 226–231, 227*f*
    discovery, 213–218
    freshwater availability, 226–231
    Polynesian cultivation, 225, 226*t*
    settlement, 213–218, 217*t*
    spatiotemporal deforestation patterns,
      221–225, 222–223*f*
  proposals of prehistoric climate change,
    94–98
  soils, 11–12, 11*t*
  tectonic setting of, 3, 4*f*
  topographic sketch map, 1–2, 2*f*
  vascular flora, 14–19
  vegetation, 19–25, 20*f*
  volcanoes of, 3–5
Easter Island collapse theory, 159
Easter Island Radiocarbon Ages (EIRA),
    107–109
  age-class frequency histograms, 108*f*
  graphical display, 109*f*
Easter Island Statue Project Website, 33
Easter Seamount Chain (ESC), 2–3
East Pacific Rise (EPR), 3
Ecocidal theory, 75–76
Ecocide, 85
El Niño Southern Oscillation (ENSO), 7–8
ENSO activity, 94–95, 96*f*
ENSO variability, 97–98, 98*f*
Environmental determinism, 235–237
Environmental-Human-Landscape
    Feedbacks and Synergies (EHLFS)
    approach
  components, 233–234
  in Easter Island, 239–244, 240–242*f*
  environmental determinism, 235–237
  general considerations, 233–238
  holistic approach, 233–234, 234*f*, 242*f*
  human determinism, 237
  human influence, 235
  multiple working hypotheses (MWH),
    238–239
  pristine condition, 235, 236*f*
  strong inference (SI), 238–239
Ephemeral settlement, 49
Eroded soils, 11–12

Ethnography, 42–44
European explorers, 42–44
Exotic species, 25
Expansion Period/Ahu Moai Period, 44–46
Extinct species, 19

F

Father Sebastian Englert Anthropological
    Museum, 30–33
Ferns, 139
First settlers, 49–60
    from Polynesia to America, 55–59
Forest clearing, 147–148
Fourier transform infrared spectroscopy
    (FTIR), 192
Freshwater availability, 226–231
Fundo Vaitea, 23–24

G

Genocide, 42–44, 83–85
Geo-climatic model, 95–97
Geologic map, Easter Island, 3–5, 5*f*
Geology, 1–5
Glacial-interglacial cycles, 91–92, 93*f*
Glacial landscape, 91–92, 92*f*
Golden Age, of Rapanui society, 46–47
Groundwater system, 8–11
Gunz glaciation, 91–92, 93*f*
Gymnosperms, 12

H

Hare paenga, 28–30, 29*f*
Heyerdahl hypothesis, 50–51, 55, 56*f*, 59
Historical trends, 42–44, 43*f*
Hornworts, 13
Hotu Iti/Mata Iti, 60–62
Human determinism, 237
Huri Moai of Decadent Phase, 75–76
Huri Moai phase, 44–46
Hydrological model, 8–11, 10*f*
Hydrosere, 138–139
Hypothetical cause-effect model,
    147–148, 148*f*

I

Impasse, 177
In situ irrigation methods, 242–243
INTCAL98, 167–168

Intermediate phase. *See* Transitional phase
Intermittent settlement, 49
Intertropical Convergence Zone (ITCZ), 7*f*
Ipu kaha/bottle gourd, 21–23, 22*f*

K

KAO1 core, 119*t*
    age-depth model, 140*f*
    pollen diagram, 141*f*
    radiocarbon-dated peat and lake
        sediment, 102–105*t*
    total deforestation, 167
KAO2 core
    age-depth model, 167–168, 169*f*
    pollen diagram of, 166*f*, 167
    radiocarbon dating, 165–167, 166*f*
Kao volcano, 3–5, 5*f*

L

Land distribution, prehistoric Rapanui,
    60–62, 61*f*
Land planning map, 34, 35*f*
Landscape deterioration, 25–27
Landscape opening (LO), 208–210
Land use, 12
Lapita, 51–52
Last Glacial Maximum (LGM), 94–95, 135
Late Glacial layer, 192–194
Late Period, 44–46
Late Pleistocene dates, 107–109
LIA drought, 220, 226–228
Lichens, 12–13
Lithic mulching, 76–78
Lithostratigraphy, 102–105*t*, 193*f*
Little Climatic Optimum (LCO), 95–97
Little Ice Age (LIA), 93–95, 183–184
Liverworts, 13, 13*t*
Livingstone corer, 102–105*t*, 174
Long-distance dispersal hypothesis,
    55–57, 56*f*
Lotka-Volterra predator-prey methods,
    81–82

M

Magnetic susceptibility, 184–186
Make Make, 63–64, 67–68
Manavai, 28–30, 76–78, 78*f*
Manutara (sooty tern), 67–68, 69*f*

Mathusian methods, 81–82
MCA drought, 219–220, 226–228
Meadow expansion, 214–215
Medieval Climatic Anomaly (MCA), 93,
    158–159, 183–184
Medieval Warm Period (MWP), 95–98
Mediterranean biome, 237–238
Microfossils, 201–203, 218
Middle Period, 44–46
Mindel glaciation, 91–92, 93f
Miniature universe, 74–75
Miru clan, 62
Moai, 34
    affected by surface weathering, 37, 38f
    fragmentation, 44–46, 45f
    water-repellent treatment, 37–38
Moai cult, 60, 63–67, 242–243
    Ahu Nau Nau, 63–64, 63f
    transport techniques, 64–65, 66f
    unfinished statues, in Raraku quarry,
        66–67, 67f
Moss species, 13, 14t
Multiple working hypotheses (MWH),
    238–239

N
National Forestry Commission (CONAF),
    34, 39
Newcomers hypothesis, 55, 56f
Nonvascular plants, 12
NUTA code, 114–115
NZA code, 114–115

O
Obsidian flaked tools (mata'a), 75–76, 77f
Obsidian hydration dating, 68–69, 70f, 79
ONF International, 39
Orongo
    ceremonial village of, 68–69, 70f
    vascular flora in, 14–16
Oxygen isotope ratio ($\delta^{18}O$), 98f

P
Paleoclimatic inference, 156–157
Paleoecological implications, 156–157
Paleoecological research, phases of,
    106–107, 107f
Paleoecological synthesis, 136–149

human impact, 147–148
paleoclimatic inferences, 145–147, 146f
paleoenvironmental inferences, 145–147
vegetation dynamics, 137–145
Paleoecology
    before, 134–135
    chronology and sedimentary patterns,
        107–112
    definition of, 90–91
    on Easter Island, 93–94
        cores retrieved and proxies studied,
            101–107, 106f
        proposals of prehistoric climate
            change, 94–98
    general considerations, 90–92
    sedimentary features of
        Rano Aroi, 99
        Rano Kao, 99–100
        Rano Raraku, 100–101
Palm phytoliths, 174, 197–199
Palm pollen, 135–136
Pavel's refrigerator technique, 64–65, 66f
Permanent settlement, 49
Petroglyphs, 28–30, 30f
    in Birdman cult, 68–69, 70f
Physicochemical proxies, 218
Phytolith analysis, 174
Pioneer phase, 106–107
Poike, 3–5, 5f
Pollen
    analysis, 135–136
    types, 159, 160–164t
Pollen diagram
    KAO1 core, 141f
    KAO2 core, 166f, 167
Pollen-vegetation relationships, 26–27
Polynesian crops, 192–203
Polynesian cultivation, 225
Polynesian-introduced flora, conservation
    of, 39–40
Polynesians, 213–214
Poz code, 114–115
Prehistoric climate change, proposals of,
    94–98
Prehistory
    ancient Rapanui society, 60
        Birdman cult, 67–72
        collapse/resilience, 74–80

deforestation, 72–74
demography, 80–82, 81*t*
genocide, 83–85
moai cult, 63–67
sociopolitical organization, 60–62
chronological summary, 85, 86*f*
chronology, 44–48
exploration to research, 42–44
first settlers, 49–60
Primeval soils, 171–172, 172*f*
Pteridophytes, 12, 14–16, 15*t*
Puringa, 214–215

Q
Quaternary glacial history, 91–92
Quaternary paleoecology, 90–91

R
Radiocarbon dating, 68–69, 71*f*, 135, 199
from cores, 115
of KAO2 core, 165–167, 166*f*
of RA2 core, 174, 175*f*
Rano, 8–11
Rano Aroi
age-depth model, 142–143, 142*f*
aquatic vegetation, 99
Late Pleistocene dates, 107–109
MCA drought in, 219–220
paleoclimatic interpretation, 145, 146*f*
radiocarbon ages from cores, 115–119*t*
sedimentary features, 99
sediments, 145–147
spatiotemporal deforestation patterns,
221, 223*f*
transitional phase, 174
UWITEC platform at, 101*f*, 102–105*t*
Rano Aroi Iti core (RAI), 199
Rano Kao
age-depth model, 139–141, 140*f*,
184–186, 185*f*, 187*f*
charcoal fragments, 214
coring intensification and reanalysis,
184–189, 185*f*, 187*f*
cultural consequences, 227–228, 227*f*
Late Holocene, pollen diagram for,
194–196, 195*f*
pollen diagram of core KAO2, 188*f*,
189–191

pollen signal, 214–215
radiocarbon ages from cores, 119–125*t*
Russian corer at, 101*f*, 102–105*t*
sedimentary features, 99–100
sediments, 214
seed/fruit dating, 184–186
semiaquatic species, 99–100
spatiotemporal deforestation patterns,
221–223, 223*f*
transitional phase, 165–168
Rano Raraku
age-depth model, 111–112, 111*f*,
137–138, 137*f*, 180–181, 181*f*
aquatic vegetation, 100–101
coring intensification and reanalysis,
180–184, 181–182*f*, 184*f*
forest clearing, 181–183
MCA drought in, 219–220
Middle Holocene, 107–109
paleoclimatic interpretation, 145, 146*f*
pollen analysis, 181–183, 182*f*
radiocarbon ages from cores, 125–131*t*
sedimentary features, 100–101
sedimentary gap, 183–184
soil erosion, 145–147
spatiotemporal deforestation patterns,
221, 222–223*f*, 223–224
transitional phase
paleolimnology and Amerindian
influence, 168–171
soil and vegetation degradation,
171–173
UWITEC platform at, 101*f*, 102–105*t*
Rapa Nui, 2–3
Rapa Nui Archaeological Database, 33
Rapanui cultivation, 225, 226*t*
Rapanui culture, 242–243
Rapa Nui Interactive Radiocarbon
Database, 33
Rapa Nui National Park, 33–40
UNESCO criteria, 33–34
Rapanui society. *See also* Ancient Rapanui
society
genocide of, 42–44
"Golden Age" of, 46–47
sweet potato and, 55
Raraku cone, 3–5
Raraku core, 205

Raraku quarry, unfinished moai statues in, 66–67, 67*f*
Raraku sediments
   internal architecture, 111–112, 111*f*
   sedimentological report, 100–101
Refrigerator technique, Pavel's, 64–65, 66*f*
Revival phase, 106–107, 179
Riss glaciation, 91–92, 93*f*
Rock art, 28–30
Roller method, for transport, 64–65, 66*f*
Rongorongo script, 30–33, 32*f*
Russian corer, 101*f*, 102–105*t*

S

Salas y Gómez Island, 3
Sea surface temperature (SST), 8
Sedimentary gap, 183–184
Sedimentary patterns, 107–112
Settlement, 213–218
   definition of, 49
SHCal13, 107–109, 114
Shepardson Moai Database, 33
Shepardson's website, 33
Smallpox, 84–85, 84*f*
Socioecological synthesis, 149–154, 159
Sociopolitical organization, 60–62
Soils, 11–12
   eroded, 11–12
   physicochemical parameters, measurement, 11–12, 11*t*
   primeval, 171–172, 172*f*
Southern Oscillation Index (SOI), 8
South Pacific Anticyclone (SPA), 7–8
South Pacific Convergence Zone (SPCZ), 7–8, 7*f*
Spatiotemporal deforestation patterns, 221–225, 222–223*f*
SRR code, 114–115
Strong Inference (SI) frameworks, 238–239
Subaquatic macrophytes, 138–139
Sweet potato, 21–23, 22*f*
   anthropogenic dispersal of, 55–57, 58*f*
   hypotheses for arrival, 55, 56*f*

T

Terevaka, 3–5, 5*f*
Terracing techniques, 242–243
Terrestrial glacial record, 91–92
Topographic sketch map, 1–2, 2*f*
Transitional phase, 106–107, 165
   Rano Aroi, 174
   Rano Kao, 165–168
   Rano Raraku
      paleolimnology and Amerindian influence, 168–171
      soil and vegetation degradation, 171–173
Transport, moai cult and, 64–65, 66*f*
Tree plantations, 23–24, 24*f*
Tuamotu Islands, 51*f*, 52–54
Tuberculosis, 84–85, 84*f*
Tuu/Mata Nui, 60–62

U

Umanga mo te Natura, 39
Universite Laval (ULA) code, 114–115
Upper altitudinal forest limit, 157–158
UWITEC platform, 101*f*, 102–105*t*

V

Vascular flora, 14–19
Vegetation, 19–25
   dynamics, 137–145
Volcanic eruptions, 220
Volcanoes, of Easter Island, 3–5

W

Warfare, 44–46, 45*f*
Water-repellent treatment, moai, 37–38
W-E navigation, 51–52
William Mulloy typing field notes, 74–75, 75*f*
Wooden sledge method, 66*f*
Woody plant taxa, 176*t*, 177
World Monuments Fund (WMF), 36–37
Wurm glaciation, 91–92, 93*f*